The Institutional Dimensions of
Environmental Change

Global Environmental Accord: Strategies for Sustainability and
Institutional Innovation
Nazli Choucri, editor

The Institutional Dimensions of Environmental Change
Fit, Interplay, and Scale

Oran R. Young

A volume prepared under the auspices of the research program on the Institutional Dimensions of Global Environmental Change (IDGEC), a core project of the International Human Dimensions Programme on Global Environmental Change.

The MIT Press
Cambridge, Massachusetts
London, England

This book was set in Sabon by Achorn Graphic Services, Inc., and was printed and bound in the United States of America.

Library of Congress Cataloging-in-Publication Data

Young, Oran R.
 The institutional dimensions of environmental change : fit, interplay, and scale / Oran R. Young.
 p. cm.—(Global environmental accord, strategies for sustainability and institutional innovation)
 "A volume prepared under the auspices of the research program on the Institutional Dimensions of Global Environmental Change (IDGEC), a core project of the International Human Dimensions Programme on Global Environmental Change."
 Includes bibliographical references and index.
 ISBN 0-262-24043-2 (hc. : alk. paper)—ISBN 0-262-74024-9 (pbk. : alk. paper)
 1. Global environmental change. 2. Environmental policy—International cooperation. I. Title. II. Global environmental accords.

GE149.Y68   2002
363.7'0526—dc21

                                                          2001056231

To all the members of the IDGEC team, with appreciation

Contents

Acknowledgments

It is a pleasure to record several major debts incurred in the course of preparing this book. Although I have received much appreciated intellectual encouragement and material support from many quarters, two sources of assistance stand out above all the others. I learned a great deal from those who have worked with me from 1994 onward to develop the scientific framework for the large-scale project on the Institutional Dimensions of Global Environmental Change (IDGEC) and to launch this project under the auspices of the International Human Dimensions Programme on Global Environmental Change. I am indebted especially to those who served first on the IDGEC Scientific Planning Committee and more recently on the project's Scientific Steering Committee. They bolstered my enthusiasm for the work involved in preparing this book. In a number of cases, they saved me from errors regarding specific issues. I hope I was able to contribute as much to their thinking as they contributed to mine.

I owe a major debt of gratitude as well to the Centre for Advanced Study in Oslo, Norway, where I enjoyed a splendid year as a fellow during 1999–2000 and where I was able to prepare substantial drafts of all the chapters of this book. The Centre proved to be an ideal environment for this type of writing, and I profited enormously from daily interactions with my colleagues in the Centre's project on the effectiveness of international environmental regimes. Arild Underdal, coordinator of our project, was particularly helpful in both intellectual and organizational terms. I can only hope that my successors as fellows of the Centre will find their time there as stimulating as I found mine.

Alf Håkon Hoel of the University of Tromsø and Ronald Mitchell of the University of Oregon, who were not members of the Centre project but with whom I interacted regularly during my year in Norway, read the entire manuscript and made extensive comments for which I am deeply grateful. Three reviewers for the MIT Press read the manuscript with care and provided me with a large number of detailed comments. One of the three, M. J. Peterson, not only identified herself but also helped me substantially to clarify matters regarding the book's audience and the nature of the links among its three substantive sections. I am grateful to these readers for helping me to correct a number of small but by no means trivial mistakes in the penultimate draft of the manuscript.

Several individual chapters appeared elsewhere in somewhat different forms. The inaugural issue of the new journal *International Environmental Agreements* carried an earlier version of chapter 2. I prepared a somewhat different version of chapter 4 for a project organized by the Committee on the Human Dimensions of Global Change of the U.S. National Academy of Sciences. It is included as a chapter in the resultant book entitled *The Drama of the Commons: Institutions for Managing the Commons.* A shortened version of chapter 6 appeared in French in a symposium on the environment and international relations published in the journal *Critique internationale.* I am grateful to the publishers in each case for permission to use these materials here.

Introduction

The members of the large and growing community of researchers interested in the roles that institutions play both in causing and in confronting various types of environmental changes have much in common. For the most part, they subscribe to the tenets of the movement in the social sciences widely known as the "new institutionalism" (Powell and DiMaggio 1991; Rutherford 1994; Scott 1995). They assume that social institutions constitute a potent driving force, accounting for a significant proportion of variance in the condition of many biogeophysical systems, and looming larger and larger as we move toward a world of human-dominated ecosystems (Turner et al. 1990; Vitousek et al. 1997). They are dedicated as well to empirical research on institutions as they actually operate or, in other words, on systems of rules in use in contrast to the study of formal rules or rules on paper, which was characteristic of an earlier generation of research on institutions (Ostrom 1990).

These are strong bonds, and they clearly provide a common direction to the work of those concerned with the institutional dimensions of environmental change. Even so, it is hard to detect a steady growth of consensual and cumulative knowledge in the work of this community of scholars. There is a clear sense that the members of this group are working on similar or parallel issues, such as conditions governing the formation of environmental regimes or factors determining the effectiveness of these arrangements once they are put in place. But on closer inspection, it often turns out that individual studies do not have enough in common to make it possible to compare and contrast their findings rigorously. The result is a proliferation of individual findings that are tantalizingly similar but fail to yield a core of agreed-upon propositions.

What is the source of this problem? No doubt, a number of factors are relevant. But in my judgment, the essential source is inability or unwillingness on the part of researchers concerned with the institutional dimensions of environmental change to adopt uniform definitions of central concepts, to specify key variables in a fully compatible manner, and to make use of harmonized data sets in evaluating major hypotheses. To make this point more concrete, take the case of institutional effectiveness construed as a dependent variable, and consider the efforts of those who approach this variable in terms of compliance with rules, behavioral change, problem solving, or movement toward some collective or social optimum. Analysts who work with these different perspectives have much in common. They all want to know whether institutions matter and how much of the variance in outcomes can be attributed convincingly to the operation of institutions. Yet there is no straightforward way to compare and contrast their conclusions about the relative importance of matters such as problem structure, actor attributes, and institutional features as determinants of the effectiveness of environmental regimes.

What is to be done about this? I believe the solution lies in striking a balance between the development of a common structure and identification and preservation of personal niches that appeal to individual researchers. Those who study the growth, operation, and effects of institutions will not and should not allow themselves to be strait-jacketed by a top-down effort to dictate research foci and strategies. Ample scope must be available for individual creativity and bottom-up initiatives to maintain vibrancy. As things stand, this is one of the strengths of the new institutionalism. At the same time, progress in this field in the form of the development of cumulative knowledge requires practitioners to accept a common structure with regard to definitions, variables, and data sets. This is an area in which the new institutionalism is relatively weak.

The goal of this book is to address this complex of issues in a manner that will speak to the concerns both of members of the research community seeking to shed light on the institutional dimensions of environmental change, and of members of the policy community in need of improved knowledge regarding institutional issues. In pursuing this goal, I take the research program of the long-term, international project on the Institutional Dimensions of Global Environmental Change (IDGEC) as a point of departure. Specifically, I treat the analytic themes identified in the IDGEC Science Plan (Young et al. 1999) as cutting-edge concerns for research on the institutional dimensions of environmental change and endeavor to develop them conceptually and analytically in a manner that can provide a common structure for individuals and groups working in this area. Not everyone will agree with the research agenda set forth in the IDGEC Science Plan, and it is certainly true that this plan does not exhaust the range of cutting-edge concerns of interest to students of institutions. But the problem I identify is real, and I have reason to believe that many members of the research community concerned with environmental change are more receptive to the development of this sort of common structure today than they were in the past.

Of interest, the need to strike a balance between common structure and personal niches is largely taken for granted among natural scientists who are socialized from an early stage in training to accept the discipline required to produce cumulative knowledge. In fact, this sort of socialization is so effective that natural scientists seldom debate such matters and often are not even conscious of their participation in joint or collaborative endeavors. But this is not the case in the social sciences, which have long been the province of creative individuals and which tend to reward those who devote themselves to formulating effective critiques of conceptual or analytic frameworks of others and seeking to replace existing frameworks (or paradigms) with new ones of their own making. No doubt this streak of individualism will continue to operate during the foreseeable future. Yet it seems fair to say that awareness is growing, at least among some subgroups of social scientists, of the limits that this mode of operation imposes on the results they can achieve. Partly, this is a matter of inner-directed feelings of frustration among those seeking to contribute to the handling of pressing public concerns. In part, it is a matter of other-directed desires to form partnerships with natural scientists who are working on related issues and who take willingness to accept common structures for granted. Nowhere is this more apparent than in the analysis of large-scale environmental changes where both of these reasons to think hard about the balance between common structure and personal niches are clearly in play.

The organization of this effort to devise a common structure to guide the efforts of those working on the institutional dimensions of environmental change is straightforward. In this book I proceed in three linked steps. The first section, encompassing chapters 1 and 2, deals with a range of conceptual issues and with construction of models designed to illuminate the causal roles that institutions play. Chapter 1 introduces the idea of the institutional dimensions of environmental change, identifies the principal science questions that arise, and describes major analytic challenges facing those seeking to answer these questions. The chapter explicates the general structure of the IDGEC Science Plan, explains the rationale underlying research priorities developed in the plan, and attempts to build a solid analytic foundation for substantive projects carried out in this common structure.

Chapter 2 turns to a central puzzle that members of the research community must address to make progress in understanding the institutional dimensions of environmental change. How can institutions, which are not actors in their own right, influence the course of human-environment relations? Is it possible to amalgamate or at least to link the perspectives of those who look at institutions through the lenses of economics, public choice, and decision theory and think in terms of collective-action models with the perspectives of those whose thinking is rooted in anthropology, ethnography, and sociology and whose ideas are expressed in social-practice models? Real progress in this field will require persuasive answers to these questions.

The next section, which contains chapters 3 through 6 and forms the heart of the book, addresses cutting-edge themes or analytic frontiers identified as research priorities in the IDGEC Science Plan. Although there is always room for innovation in framing research agendas, these themes capture the evolution in thinking about institutions from the focus of the 1990s on matters of regime formation and effectiveness to growing emphasis on comparing the operation of institutions at different levels of social organization and on examining interactions among distinct or discrete institutional arrangements. In the terminology of IDGEC, these themes are known as the problems of fit, interplay, and scale. They center, respectively, on the (mis)match between properties of biogeophysical systems and attributes of institutions, on interactions between and among distinct institutions, and on the prospects for scaling up or down in the dimensions of space and time in our efforts to understand the roles that institutions play in causing and confronting environmental change.

Chapter 7, which makes up the third section, turns to the links between analysis and praxis. Because most of those working on the institutional dimensions of environmental change are interested in solving problems as well as in generating new knowledge, this section should be of interest to members of the research and policy communities alike. It develops a perspective on the links between analysis and praxis that differs from the view of a sizable proportion of contributors to the new institutionalism. It introduces the idea of usable knowledge and discusses alternative strategies for applying knowledge about the roles that institutions play to concerns of those in the policy community who are seeking effective ways to avoid or mitigate environmental changes, or to adapt to such changes once it becomes apparent that they are unavoidable. I offer a critique of the idea of design principles as a way of organizing thinking about the issue of usable knowledge (Ostrom 1990) and propose a procedure known as institutional diagnostics as a means of bridging the gap between science and policy in this realm.

Throughout the book I turn to large-scale environmental changes, such as depletion of stratospheric ozone, climate change, degradation of large marine and terrestrial ecosystems, and losses of biological diversity, to illustrate my points. The result is particular interest in the role of international institutions as sources of environmental change and prospects for establishing international regimes that will prove effective as mechanisms for solving or ameliorating the resultant problems. But the roles that institutions play are not limited to international society. On the contrary, institutions loom large in accounts of environmental changes occurring at all levels of social organization. Even global environmental changes are often products of large numbers of small-scale actions (e.g., actions that degrade critical habitat) or interactions between human actions at the local or subnational level (e.g., combustion of fossil fuels) and the behavior of global ecosystems (e.g., Earth's climate system). As a result, any study must take into account human actions occurring at many levels of social organization. The knowledge produced should prove interesting to all those interested in the role of institutions in this field regardless of where they are located in the overall social order.

I

Concepts and Models

1

Environmental Change: Institutional Drivers, Institutional Responses

Institutions figure prominently in most accounts of the causes of major changes in biogeophysical systems as well as in many prescriptions for solving problems arising from these changes or, more modestly, ameliorating their effects on human welfare. Many observers regard unsustainable uses of renewable resources, such as stocks of fish and wildlife, grazing lands, forests, soils, and groundwater, as consequences of systems of property or use rights that fail to give human users adequate incentives to devote energy and resources to conserving these resources. These observers also explain emissions of harmful pollutants, including sulfur dioxide and nitrogen oxides, ozone-depleting substances, greenhouse gases, and persistent organic pollutants, in terms of the absence of regulatory rules necessary to provide owners or users of various factors of production with effective incentives to avoid or minimize social costs arising as unintended byproducts of their actions.

Conversely, strategies aimed at coming to terms with environmental problems frequently call for changes in prevailing structures of property rights, introduction of new regulatory regimes, or development of appropriate incentive mechanisms (e.g., charges or tradable permits) as procedures for redirecting or guiding the behavior of those whose actions lead to anthropogenic disturbances of ecosystems. Suitably adapted to the circumstances of specific problems, recommendations may be directed toward actors ranging from local appropriators of fish or groundwater to operators of large power plants that are major emitters of greenhouse gases.

Emergence of sustained interest in what may be called the institutional dimensions of environmental change is therefore easy to understand. Yet it is worth noting at the outset that this theme differs in at least two fundamental ways from most other topics that have emerged as major research foci. Emphasis on institutions directs attention to a particular suite of independent variables or, in terminology that has become familiar in this field, driving forces in contrast to a specific type of environmental change, such as depletion of fish stocks, loss of biological diversity, or thinning of the stratospheric ozone layer. Those concerned with phenomena such as changes in patterns of land use necessary to protect biological diversity or transformation of industrial systems to reduce emissions of greenhouse gases concentrate on key dependent variables and consider the role of a variety of driving forces. Analysts interested in contributions of institutions seek to understand the roles that institutional drivers play in a range of environmental changes. Thus, the study of institutions constitutes a cross-cutting theme for those interested in what has become known as global environmental change (National Research Council 1999c).

Institutions play more or less significant causal roles with regard to most environmental changes involving human action. Yet—and this is the second distinctive feature of this research agenda—institutions seldom account for all of the variance in these situations. In the typical case, they are one among a number of driving forces whose operation, both individually and in combination, generates relevant environmental changes. And it is reasonable to expect that substantial variations will occur in the significance of roles that institutions play from one situation to another. A prominent feature of research, therefore, is a sustained effort to separate the signals associated with institutional drivers from those associated with other drivers, and to understand how different driving forces interact with each other to account for observed outcomes.

The Focus on Institutions

Over the last two decades, a movement known to many as the new institutionalism has gathered force throughout the social sciences. Several common concerns animate this movement in all its manifestations. The new institutionalism is pragmatic, empirical, and marked by emphasis on "rules in use," in contrast to formal provisions of contracts, constitutions, treaties, or other constitutive documents (Ostrom 1990). What holds the movement together and differentiates it from older forms of institutionalism (Powell and DiMaggio 1991; Rutherford 1994) is a desire to understand the actual roles that institutions play as determinants of the outcomes of interactive human behavior, or the links between micromotives and macrobehavior in various social settings (Schelling 1978). This does not imply lack of interest in institutional design. But from this perspective, design makes sense only to the extent that it involves understanding of the ways institutions are likely to work in practice. This is in contrast to a study of outcomes that would occur in an idealized world in which those subject to the rules and procedures of institutions internalize them and comply fully with their requirements under all circumstances.

The new institutionalism offers a broad umbrella that shelters a range of perspectives on human affairs. Most contributors are comfortable with a point of departure that treats institutions as sets of rules, decision-making procedures, and programs that define social practices, assign roles to the participants in these practices, and guide interactions among the occupants of individual roles. Structures of property rights, electoral systems, and practices relating to marriage and the family are all examples of institutions in this sense. Institutions, on this account, must not be confused with organizations construed as material entities with employees, offices, equipment, budgets, and (often) legal personality. In rough and ready terms, organizations (e.g., Exxon Corporation, the U.S. Republican Party, the World Bank) can be thought of as actors that typically emerge as players whose activities are guided by the rules of the game of institutions in which they participate. Conceptualized in this way, institutions can and do vary widely in terms of a range of dimensions, including functional scope, spatial domain, degree of formalization, stage of development, and interactions with other institutions. Institutions that deal explicitly with environmental or resource issues are commonly known as environmental or resource regimes (Young 1982).

Beyond this, the perspectives of those whose thinking about institutions is rooted in disciplines as diverse as economics and anthropology diverge substantially. To begin with, an important distinction exists between thin perspectives and thick perspectives on institutions. Institutions in the thin sense are systems of rules, decision-making procedures, and programs as

articulated in constitutive documents (e.g., contracts, constitutions, treaties). This is the normal point of departure for those who speak of the
rules of the game as defining features of institutions and leave open as
suitable topics for analysis all questions regarding behavioral consequences of these arrangements (North 1990). Institutions in the thick
sense are social practices that are based on the rules of the game but also
include common discourses in terms of which to address the issues at
stake, informal understandings regarding appropriate behavior on the
part of participants, and routine activities that grow up in conjunction
with efforts to implement the rules (Scott 1995). Although they are typically based on rules, procedures, and programs articulated in constitutive
agreements, social practices ordinarily evolve over time in ways that are
not easy to trace to their constitutive foundations, even though they are
likely to be well understood by participants. This is part of the message
associated with the familiar contrast between rules in use and rules on
paper. Among other things, this distinction provides the point of departure for debates occurring in many settings between strict constructionists, who take the formulations set forth in constitutive documents as
their guide and endeavor to bring behavior into conformity with them,
and liberals, who are content to adjust or even reinterpret formal characterizations of constitutive provisions to bring them into line with changing circumstances.

What is the significance of this distinction between thin and thick definitions of institutions? The two conceptions ordinarily identify universes
of cases that overlap but are by no means identical. The thin definition
counts all sets of rules and procedures articulated in constitutive agreements as institutions, regardless of their behavioral significance. The thick
definition treats behavioral consequences as a defining characteristic of
institutions. It omits dead letters from the universe of cases (Rittberger
1993) and at the same time includes de facto practices that do not rest
on formal constitutive agreements. The two conceptions are also likely
to produce descriptions of specific institutions that are not fully congru-
ent, even when they agree regarding the inclusion of these arrangements
in the same universe of cases. This is a consequence of the gap between
rules on paper, highlighted by the thin definition, and rules in use, emphasized by the thick definition. It follows that differences will occur as well

regarding perspectives on effectiveness embedded in the definitions and, as a result, in the views of those making use of these definitions to examine the determinants of effectiveness. Simply put, the thin definition directs attention to matters of compliance or conformance, whereas the thick definition focuses on a broader range of behavioral patterns arising in conjunction with the operation of social practices.

Whether we adopt a thin perspective or a thick one, it is helpful to draw a distinction between the strength or depth of institutions on the one hand and their robustness or durability on the other. Strength is a measure of the extent and stringency of an institution's rules and practices; that is, the extent to which the institution requires subjects to alter or adapt their behavior to conform to its requirements. Those who think in these terms often posit links between strength and compliance. They argue, for instance, that the deeper (i.e., stronger) cooperation becomes, the harder it will be to achieve high levels of compliance on the part of individual members (Downs, Rocke, and Barsoom 1996). But note that a strong institution whose members have only mediocre records of compliance may actually prove more significant in functional terms than a weak institution whose members always comply with its undemanding requirements.

Robustness, in contrast, is a measure of the capacity of an institution to survive various pressures intact in the sense of withstanding the impact of destabilizing forces without suffering collapse or experiencing transformative change (Hasenclever, Mayer, and Rittberger 1999). Empirical studies of robustness are hampered by difficulties encountered in devising operational measures of collapse or transformation, but several significant conceptual distinctions are worthy of note. Destabilizing forces may be either endogenous (e.g., the operation of democratic electoral systems can lead to selection of antidemocratic leaders) or exogenous (e.g., revolutionary changes in overarching political systems can overwhelm more specific arrangements governing human uses of renewable resources). Similarly, such forces may take the form of sharp shocks or crises (e.g., sudden collapses in fish stocks) or pressures whose intensity increases or builds more gradually (e.g., rising sea levels). Institutions that have considerable capacity to adjust to pressures that increase gradually may succumb to severe shocks (e.g., monetary crises) almost overnight. In some cases, institutions whose members are able to mobilize effective responses to sudden crises have little capacity to adjust to pressures that rise slowly—sometimes even imperceptibly—but are more systemic. Robustness, then, is a multidimensional variable, a fact that is likely to lead to the formulation of a number of hypotheses intended to explain why some institutions are more durable than others.

Another conceptual matter of obvious relevance concerns the density of institutions operating at the same time within a given social space (e.g., local society, national society, international society). Identifying clear boundaries may not be straightforward in specific cases. Like ecosystems, individual institutions are often linked together through various types of interdependencies (Commoner 1972). Even so, every society encompasses a number of institutions separated in terms of functional scope, spatial domain, or membership. The density of institutions operative in specific social settings is a variable. Most national societies feature a high density of institutions. International society, in contrast, has long been treated as a low-density setting. But there are clear indications that the density of distinct institutional arrangements operating at the international level has been on the rise for some time, especially with regard to functional concerns such as economic relations and the environment.

Density is a matter of interest both as a dependent and an independent variable. For example, considerable interest is expressed in finding ways to measure trends in the density of international institutions and to identify conditions that account for increased density in this social space (Meyer et al. 1997). Although this development is a matter of interest in its own right, the effort to explain these trends is driven in considerable part by interest in the likely consequences of increases in density of international institutions. Some observers suggest that increasing density will lower the probability of severe (even violent) conflict at the international level on the grounds that the cost of conflict rises as a function of the number and variety of institutional arrangements it is likely to disrupt (Conca 2001). In addition, rising density leads to an increase in interactions between or among distinct institutions. Such institutional interplay is a familiar phenomenon at the domestic level where institutional density has been high for some time; numerous procedures have evolved over time to deal with these interactions in ways that protect or enhance social welfare. But rising levels of institutional interplay are a more novel concern at the international level. Finding ways to handle them is especially challenging in a social setting that lacks a central public authority—or what we normally think of as a government—authorized to deal with such matters.

These comments lead to the observation that institutions created to deal with specific environmental problems or issues frequently become embedded in larger hierarchical structures (von Moltke 1997). Consider developments associated with exclusive economic zones (EEZs) as a case in point. Created during the 1970s and 1980s and formalized in the 1982 U.N. Convention on the Law of the Sea (UNCLOS), EEZs cover about 8 percent of the Earth's surface, an area encompassing some 25 percent of global primary productivity and 90 percent of the world's fish catch (Independent World Commission on the Oceans 1998). This shift in the rules of the game is one of the most important changes of modern times in institutional arrangements prevailing in international society. But EEZs do not operate as isolated or stand-alone arrangements. They provide a framework for individual countries to devise national regimes dealing with fish and other resources located within the zones. In some cases, subnational and even local arrangements deal with the resources in question. This has led some commentators to use the term meta-regime in seeking to understand the significance of EEZs as a major institutional innovation in international society. What is more, EEZs themselves form a part of the overarching arrangements for the world's oceans codified in the UNCLOS, and they are embedded in broader institutional arrangements that make up the deep structure of international society.

Any effort to determine the impact of EEZs will therefore require assessment of the larger institutional structures of which they form integral elements. In all likelihood, this will lead to mixed conclusions. The results flowing from the introduction of EEZs are likely to vary considerably as a function of the character of national and subnational arrangements established to bring this meta-regime to bear in different geopolitical settings. What is more, the consequences of the EEZs, which encompass the same elements throughout international society, will be sensitive to regional variations in the character of key ecosystems. It is not surprising, for instance, that problems have arisen in cases where individual fish stocks straddle boundaries separating the EEZs of two or more coastal states or EEZs from the high seas, in contrast to cases where stocks are wholly encompassed within individual EEZs. The significance attached to negotiations leading to the 1995 Straddling Fish Stocks Agreement indicates the importance of these problems (Stokke 2001a).[1]

Finally, it is worth highlighting at the outset one important consequence of the fact that institutions—even in the thick sense—constitute cross-cutting forces. In virtually every setting involving environmental change, institutions make up only one of a set of drivers that may include nonanthropogenic forces (e.g., changes in Earth's climate system unrelated to human interventions) as well as anthropogenic forces (e.g., population growth, technological innovation, business cycles). This has a number of implications to which I return in the next section of this chapter. But the point at this stage is that an institutional arrangement that produces good results in one setting may be an outright failure in other settings. Just as a number of different institutional arrangements may be sufficient to solve a particular problem, it is risky to assume that because a particular institution yields good results in one setting, it can be expected to perform equally well in other settings. To be specific, a management system governing harvesting of fish that works well as long as the behavior of appropriators is guided by the logic of appropriateness may fail dismally in a setting where behavior is based largely on the logic of consequences (March and Olsen 1998). Similarly, a system that does fine in the absence of sharp biogeophysical fluctuations that produce sudden crises may collapse quickly in a setting where the occurrence of such crises requires rapid responses on the part of those responsible for monitoring the status of fish stocks and adjusting total allowable catches to reflect biogeophysical changes.

The Principal Science Questions

All those seeking to understand the connections between institutions and environmental change share a general interest in the roles that institutions play both in causing and in confronting disruptions in large and important ecosystems. Yet it is possible to draw distinctions among several specific science questions that animate the efforts of members of this community (Young et al. 1999). At the most basic level lies the question of causality: How much variance in the condition of ecosystems is attributable to institutions? Next is the question of performance: Why do some institutional responses to environmental problems prove more successful than others in terms of criteria such as sustainability, efficiency, and equity? At the most applied level is the question of design: How can we structure institutions to maximize their performance? Obviously, these questions are linked to one another. Basic research on causality is motivated at least in part by desire to improve our ability to design regimes that will prove effective in solving, or at least managing, specific environmental problems and, for that matter, in meeting various standards of efficiency and equity. It is impossible to design effective institutions without some understanding of the roles that these arrangements play as driving forces in the realm of human affairs. Nonetheless, individual researchers are attracted more or less powerfully to one or another of these questions, and the paradigmatic research puzzles that come into focus in connection with each question are distinct.

The Question of Causality

In the final analysis, interest in the institutional dimensions of environmental change rests on claims about the significance of institutions in causal terms. There is no implication here that institutions account for all the variance in environmental conditions. Even when they play relatively modest causal roles, it is important to understand their specific contributions in causing and confronting environmental changes. Yet claims regarding causation involve several analytic complications or puzzles that run through efforts to add to our understanding of the significance of institutions in human affairs. Because institutions are not actors in their own right, they can only affect the outcomes of interactive decision making by influencing the behavior of those who are actors. But the pathways through which this influence occurs include a number of behavioral mechanisms that are distinct but often operate simultaneously or even interact with each other under conditions that are poorly understood (Young 1999a). What is more, spatial and temporal associations between institutions and environmental changes do not provide unambiguous evidence of significant causal connections. The apparent role of institutions both in causing environmental problems (e.g., collapse of a fish stock after increases in levels of human harvesting) and in solving them (e.g., increases in biological diversity following reforestation) may turn out to be spurious. It is perfectly possible that biogeophysical forces (e.g., changes in sea water temperatures) are the real cause of collapse of a given fish stock; the ability of key species to adapt to changing ecological conditions may account for improvements in biodiversity. Under the circumstances, it will come as no surprise that the effort to pinpoint mechanisms through which institutions are causal forces with respect to ecological conditions and to demonstrate these connections is the most fundamental challenge in this field of study. To put it bluntly, in the absence of demonstrated causal links, interest in the institutional dimensions of environmental change will fade away and eventually disappear.

In thinking about this challenge, it makes sense to start with the simplest case and work toward more complex cases. Consider a relatively well-defined ecosystem (e.g., a large marine ecosystem such as the Bering Sea) and a single regime focused on that system and composed of a set of harvesting rules dealing with seasons, gear restrictions, and catch limits (Iudicello, Weber, and Wieland 1999). The issue can be framed as follows: How much of the variance in the status or condition of the living resources of the ecosystem can be attributed—ceteris paribus—to the operation of these rules? Or to put it more concretely, to what extent does the operation of the regime determine the condition of these resources? Most efforts to answer these questions fall into two categories (Young 2001b). One strategy is to begin by asking what would have happened in the absence of the regime and then to treat the effects or consequences of the regime as the difference between the no-regime outcome and the actual outcome (Helm and Sprinz 1999; Miles at al. 2001). The other strategy tracks changes in key variables (e.g., increases or decreases in various fish stocks) occurring during the period beginning with creation of the regime and determines what proportion of these changes can be attributed convincingly to the operation of the regime (Mitchell n.d.). Each strategy has merit; neither one alone offers a simple procedure for addressing or circumventing the question of causality.

A somewhat more complex issue arises in efforts to assess the relative contributions of two or more distinct drivers as determinants of the status or condition of an ecosystem. Turning again to the marine ecosystem, we can ask how much of the variance in the condition of fish stocks over time is caused by institutional drivers (e.g., rules governing human harvesting) in contrast to biogeophysical drivers (e.g., changes in sea water temperatures or increases or decreases in populations of other organisms). The challenge is not to identify a master variable in the sense of a single driver that can account by itself for the condition of the relevant ecosystem. Rather, the objective is to separate out signals of several types of drivers and to assess the causal role that each plays in determining the status or condition of the ecosystem. Nothing in this way of framing the problem yields convincing procedures for demonstrating the causal significance of individual drivers. But this approach is both compatible with the view of institutions as cross-cutting factors and conducive to thinking about the dynamics of environmental problems in terms that are relevant to policy making.

An even more complex issue arises when institutional drivers interact with other driving forces. Consider the marine ecosystem once again. A successful regime would track fluctuations in the abundance of specific stocks of fish, raising quotas when stocks increase and lowering them when stocks decrease. But suppose that those responsible for operating the regime fail to lower quotas or even raise them in response to economic or political pressures at a time when a stock is declining. Here is a case in which a stock that is perfectly capable of recovering in the absence of anthropogenic interventions may be pushed beyond the margin of recovery by the interaction of anthropogenic and biogeophysical drivers. Recent years have brought awareness of the fact that many ecosystems do not have powerful stabilizing mechanisms that can be counted on to move them back toward some preexisting equilibrium after major disturbances (Wilson et al. 1994). Even in the absence of human interventions, ecosystems frequently undergo dramatic shifts from one state to another. Interactions between institutional and biogeophysical drivers can be expected to accentuate the effects of nonlinear dynamics in a wide range of systems (National Research Council 1996).

The preceding paragraphs clarify and highlight the question of causality rather than answer it. What can be done to generate convincing answers? At this stage it seems clear that we are unlikely to find a silver bullet in the sense of a straightforward procedure that can be counted on to sort out causal connections between ecosystems and a variety of driving forces. Controlled experiments are seldom feasible in dealing with large ecosystems and complex institutions. Although procedures featuring statistical inference are useful in some settings, small and heterogeneous universes of cases often impose severe constraints on this approach to the question of causality. Under the circumstances, it is easy to understand why those seeking answers to the question of causality regularly resort to a combination of procedures featuring in-depth case studies, analysis of behavioral pathways, the procedure based on Boolean algebra and known as qualitative comparative analysis, and even computerized simulations (Underdal and Young n.d.). None of these procedures by itself is likely to yield convincing conclusions about the roles that institutions play in the dynamics of large ecosystems. But taken together, they are capable of adding incrementally, albeit sometimes slowly, to our understanding.

The Question of Performance

Assuming that institutions make a difference in the sense that they emerge as significant driving forces in a variety of settings, it is logical to ask a range of questions pertaining to their performance. Whereas the question of causality centers on establishing the significance of institutions as driving forces, the question of performance requires specification of criteria of evaluation, followed by assessment of the extent to which actual outcomes measure up in terms of those criteria. Under the circumstances, it is easy to see that an institution may be accepted as a powerful causal force but be regarded as a failure or an underachiever in terms of standards based on criteria such as sustainability, efficiency, or equity. The question of performance makes sense only in settings where it is generally agreed that institutions are significant causal forces. But once that threshold is crossed, additional issues relating to performance come into focus.

Several conceptual distinctions will help to organize thinking about the question of performance. The concept of simple performance directs attention to results flowing from the operation of an institution that are internal in the sense that they are confined to the relevant behavioral complex, direct in the sense that they involve short causal chains, and positive in the sense that they contribute to solving identifiable problems (Young 1999a). Complex performance, in contrast, subsumes simple performance and adds to it a range of broader or more extended outcomes that occur outside the initial behavioral complex, involve longer causal chains, and encompass negative as well as positive effects (Levy, Young, and Zürn 1995). Most thinking about the question of performance has focused on efforts to assess simple performance. But it is easy to see that a regime created to solve an environmental problem can achieve high marks with regard to simple performance, while at the same time generating broader outcomes that offset its contribution and that may even produce a situation in which the net effects of the arrangement are judged to be negative from the perspective of social welfare more generally.

The most common criteria of evaluation employed by those concerned with simple performance regarding environmental problems center on the idea of sustainability in one or another of its forms. Is the fish stock robust in the sense that it produces sustainable yields over time? Is the level of biological diversity stable or even increasing? Are emissions of greenhouse gases being kept at levels that are low enough to avoid serious anthropogenic interference in Earth's climate system? Important as this biogeophysical perspective is, however, it is not the only one relevant to assessing simple performance. At a minimum, most observers would add to these sustainability considerations concern for efficiency (e.g., is sustainability being achieved in a way that minimizes costs?) and for equity (who gains and who loses, and are outcomes achieved through legitimate procedures?). Although they are distinct in analytic terms, these considerations often interact. A demonstrable commitment to the pursuit of efficiency is likely to be necessary to convince actors to agree to terms of institutional arrangements in the first place. Willingness to address matters of equity often proves critical in persuading actors to comply with requirements of institutional arrangements once they are in place.

To this set of concerns the idea of complex performance adds a range of issues that are broader and more difficult to deal with in terms of empirical analysis. An institution created to solve a well-defined problem (e.g., ensuring sustainable uses of fish stocks) may produce consequences affecting the domestic politics of individual member states (e.g., influence of the fishing industry as an interest group), other regimes dealing with issues that intersect with the arrangement at hand (e.g., rules governing international trade or investment), or social practices operating at the level of the society as a whole (e.g., practices relating to the divisibility of sovereignty at the international level). Similarly, broader outcomes that institutions produce may vary greatly in terms of directness or length of the causal chain linking the institution and the outcome. It is relatively easy, for instance, to demonstrate convincing connections with regard to immediate outcomes, such as promulgation of regulations devised for the express purpose of operationalizing rules articulated in a law or a treaty creating a specific regime. But the trail grows cold quickly as the length of the causal chain increases. This does not mean that more indirect outcomes generally considered under the heading of complex performance are insignificant or that they can be ignored safely. But it does help to account for the fact that efforts to assess complex performance are considerably less advanced than studies of simple performance.

Hovering over this discussion of the question of performance is the concept of social welfare (Underdal 1999). Even at the local level social welfare is a concept that is difficult to operationalize; complications associated with it increase proportionately as we move toward the national and international levels. Nonetheless, it is pertinent to ask whether efforts to solve environmental problems by creating institutions to regulate or govern human interventions in large ecosystems yield net improvements in social welfare. What makes this issue both interesting in substantive terms and challenging in analytic terms is the fact that social welfare in this context must take into account both the costs to individual actors arising from restrictions on their freedom to act independently, and opportunity costs arising from the fact that societies cannot use resources invested in solving one problem to deal with other problems. Even when a regime does a good job of solving a well-defined environmental problem (e.g., sustaining a fish stock, controlling intentional oil pollution at sea), critics may contend that society would be better off if resources devoted to creating and operating the governance system for that problem had been used instead to address some other problem (e.g., protecting habitat critical to endangered species). Because social agendas are typically compartmentalized and addressed in different arenas, policy makers seldom consider questions of this sort in a systematic manner. Yet a comprehensive account of the question of performance cannot afford to ignore them completely.

The Question of Design

In practice, much of the interest in the institutional dimensions of environmental change is driven by a desire to (re)design arrangements to solve more or less well-defined problems (depletion of fish stocks, thinning of the ozone layer, loss of biological diversity). Where existing institutions are treated as sources of the problem, this means modifying or replacing these arrangements to redirect the behavior of relevant human actors. Where problems are attributable to biogeophysical drivers, on the other hand, the goal is to create institutions that will give human actors proper incentives to cope with these drivers. Efforts to design specific institutions are constrained both by limitations in our ability to foresee how institutions treated as complex systems will perform in practice and by the fact that the character of institutional arrangements is more often a product of bargaining among actors pursuing their individual interests than a result of some systematic exercise in social engineering (Young 1982). Nonetheless, interest in (re)designing institutions as a means of coming to terms with environmental problems is strong and likely to become even stronger during the foreseeable future.

Because institutions are not actors in their own right, those engaged in designing them must think at all times about the probable effects of the arrangements they create on the behavior of various groups of actors. Broadly speaking, two ways of thinking dominate this endeavor: the logic of consequences and the logic of appropriateness (March and Olsen 1998). Those who think in terms of the logic of consequences assume that actors are utilitarians responding to changes that affect benefits and costs associated with available options. They will seek to design arrangements that alter incentives by driving up the costs of undesirable actions (e.g., increased harvests of fish) and increasing the benefits of desirable actions (e.g., investing in future returns from renewable resources). Those who think in terms of the logic of appropriateness assume that actors behave in ways that they regard as right or proper and that they will normally accept restrictions that they conceive of as legitimate. They will seek to design arrangements (e.g., procedures for setting allowable catches in various fisheries) that actors treat as authoritative because their voices were heard in the design process or because they are based on underlying principles that actors regard as fair or just (Risse 2000). The logic of consequences and the logic of appropriateness are not mutually exclusive; designers creating institutions that will perform well are likely to pay attention to both. Still, the two approaches do yield strikingly different outlooks on the question of design.

Beyond this lies the challenge of bringing general knowledge about causality and institutional performance to bear on design involving specific environmental problems (e.g., managing fish stocks, stabilizing Earth's climate system). One approach features the development of design principles. A design principle, as Elinor Ostrom (1990, 90) put it in her well-known analysis of common-pool resources (CPRs) is "an essential element or condition that helps to account for the success of . . . institutions in sustaining the CPRs and gaining the compliance of generation after generation of appropriators to the rules in use." This approach yields tentative generalizations spelling out necessary conditions for success of the following sort: "[m]ost individuals affected by the operational rules can participate in modifying the operational rules" and "[m]onitors, who actively audit CPR conditions and appropriator behavior, are accountable to the appropriators or are the appropriators" (Ostrom 1990, 90). The obvious implication of this exercise is that once design principles have been identified and tested through systematic empirical analysis, it should be possible to apply them in connection with efforts to (re)design institutions addressing any member of the relevant universe of cases (small-scale CPRs in Ostrom's analysis).

Needless to say, this is an appealing prospect. Yet, as Ostrom herself would be the first to acknowledge, it assumes a universe of cases that is both well defined and relatively homogeneous. As we move toward many familiar problems on today's environmental agenda, applicability of these principles becomes less and less clear. Where incentives to cheat are not strong or behavior is highly transparent, for instance, monitoring may be far less important than arrangements to enhance the capacity of actors to fulfill the commitments they have made under the terms of constitutive agreements. Where the ultimate problem centers on consumption of environmentally unfriendly goods (e.g., ozone-depleting substances), it may make sense to focus attention on the actions of the relatively small number of producers of these goods rather than on actions of the multitude of consumers. The point is not that we should eschew efforts to develop design principles. Rather, it is important to exercise great care in thinking about applying seemingly simple principles to the complexities of specific situations.

An alternative approach centers on what may be called institutional diagnostics (see chapter 7). This approach starts from the premise that one size does not fit all when it comes to designing institutions to solve environmental problems. It therefore calls for an effort to identify critical features of specific problems followed by an effort to specify institutional arrangements that are best suited to deal with the most prominent of these features in the case(s) at hand. Consider climate change. Given a significant chance that Earth's climate system will behave chaotically and generate costly surprises, the climate regime has to grant priority to creating early warning systems and procedures that allow rapid adjustments in its rules. Similarly, the fact that a relatively high level of uncertainty exists as to the nature—and even the reality—of climate change makes it important to devote resources to improving knowledge about this problem and to provide mechanisms through which social learning can lead to suitable modifications in the climate regime (Social Learning Group 2001). It is not necessary to set the search for design principles and the practice of institutional diagnostics in opposition to one another. Both may yield results that help us to cope with important environmental problems. But it is worth noting that the diagnostic approach can succeed only when a close working relationship is established between those who possess high-quality knowledge of the biogeophysical attributes of a particular problem and those who aim to devise an institutional arrangement that is best suited to the character of that problem.

Analytic Frontiers: Fit, Interplay, and Scale

Questions of causality, performance, and design define the research agenda for those interested in the institutional dimensions of environmental change. They also provide appropriate criteria of evaluation to be used in assessing the performance of members of the research community.

That much is clear. But the issue of where to cut into this subject in the interest of maximizing contributions to knowledge and helping to solve problems is more difficult to resolve. There is room for numerous strategies, and individual researchers must be allowed to make their own investment decisions in allocating their time and energy among alternative approaches to the central questions. Recently, however, researchers working on the institutional dimensions of environmental change have engaged in a coordinated effort to identify particularly promising lines of enquiry. This has led to identification of three cutting-edge themes that are known in the field as the problems of fit, interplay, and scale.

The Problem of Fit

The problem of fit centers on one fundamental proposition. An institutional arrangement that performs perfectly well dealing with one environmental problem may be a dismal failure in solving other problems. A regime based on the assumption that stocks of renewable resources are highly resilient and able to rebound quickly after a temporary ban or moratorium on harvesting will run into problems when the resources in question are easily driven to a point at which recovery is slow or even past the point of no return. Similarly, an arrangement that performs creditably in managing human activities affecting biogeophysical systems that are surprise free, in the sense that they are not subject to sudden and unexpected shifts from one state to another, will produce unsatisfactory results in settings where nonlinear changes or cascades can trigger severe crises. In essence, the problem of fit deals with congruence or compatibility between ecosystems and institutional arrangements created to manage human activities affecting these systems (Berkes and Folke 1997; Cleveland et al. 1996). Overall, the presumption is that the closer the fit between ecosystems and institutional systems, the better the relevant institutions will perform, at least in terms of sustainability.

This problem is comparatively easy to deal with when universes of cases are homogeneous. To the extent that all members of a class of environmental problems involve biogeophysical systems that are roughly comparable in terms of structures and processes, those responsible for creating regimes to manage relevant human activities can be expected, perhaps gradually, to develop a repertoire of best practices that can be applied to one case after another with reasonable expectations of success. This is not to say that knowledge regarding best practices will accumulate easily or quickly. In fact, a good deal of trial and error may be required to devise effective procedures for managing human activities even when the essential nature of environmental problems is strikingly uniform. Moreover, superficial differences may obscure the underlying similarities among different situations. Not all CPRs, for instance, look the same to a casual observer, who may be struck by the fact that they differ greatly in biogeophysical terms. Transfer of knowledge about best practices may also prove difficult in social settings where individual actors are reluctant to draw lessons from or rely on each other's experiences. Even when environmental problems are more or less the same, therefore, individual human groups may have to learn about best practices the hard way—by accumulating experience of their own rather than benefiting from experiences of others.

The problem of fit becomes more complex—and this is the important point—when the universe of cases is heterogeneous in biogeophysical or socioeconomic terms and where it may not be easy to pinpoint essential differences among individual members of the universe. Consider ozone depletion, climate change, and loss of biological diversity in these terms. All three are generally treated as prominent cases of global environmental change. But beyond this, similarities and differences among them are anything but clear. It is easy to see that ozone depletion and climate change are closely tied to industrial production, whereas loss of biodiversity is largely a matter of habitat destruction; and that ozone depletion involves a relatively small economic sector, whereas climate change and loss of biodiversity involve more fundamental economic and political arrangements. But at this stage, we lack a systematic procedure for identifying attributes of environmental problems that are most critical from a managerial perspective and analyzing their implications for developing effective institutional arrangements. Although we have good reason to believe that the universe of cases is relatively heterogeneous, the tools available for ensuring that regimes are well suited to the problems are limited.

Under the circumstances, it is not surprising that more or less serious misfits or mismatches between environmental problems and regimes are common. In the absence of systematic understanding of the problem of fit, it is tempting to proceed by analogy and especially to assume that regimes that are successful in one context will work well in other settings. Thus, many commentators are drawn to the view that we can derive lessons from experience confronting ozone depletion that will help to solve climate change, despite obvious differences between the two problems (Susskind 1994). It is also natural for individual actors to push hard for the creation of arrangements that are compatible with their own interests. Differences between members of the European Union and states that form the so-called Umbrella Group regarding matters such as carbon sequestration and emissions trading in the case of climate change, for example, are easy to understand as reflections of particular interests of the two groups. But so also are numerous other situations involving institutional preferences, such as disagreements about regulations affecting land use and standards applying to air pollution in domestic settings.

It is worth noting as well that mismatches between regimes and ecosystems are frequently difficult to eliminate. Simply pointing to poor results measured in terms of indicators of sustainability and appealing to members of the relevant group to take necessary steps to enhance social welfare is seldom sufficient to ensure positive results. Partly, this is a matter of difficulties associated with efforts to document unsustainable activities and to demonstrate conclusively the role of prevailing institutions as causes of these activities. To return to marine ecosystems, it is remarkable how hard it is to gain consensus regarding the biogeophysical status of overexploited fish stocks, much less on the role of human harvesting as a driving force in the decline or even collapse of specific stocks (Dobbs 2000). In part, the problem lies in the interests of those whose livelihoods are affected. Many fishers resist changes in regulatory arrangements even when they know perfectly well that key stocks are overexploited, because they have large debts to service on a regular basis and because they have little ability to switch from fishing to alternative ways of making a living. It follows that mismatches can be highly resistant to change, even when all parties concerned are aware that existing practices are unsustainable and generally inefficient as well.

The Problem of Interplay

Although it is tempting to treat them as self-contained arrangements, most institutions interact with other similar arrangements both horizontally and vertically. Horizontal interactions occur at the same level of social organization; vertical interplay is a result of cross-scale interactions or links involving institutions located at different levels of social organization. Interplay between or among institutions may take the form of functional interdependencies or arise as a consequence of politics of institutional design and management. Functional interdependence is a fact of life. It occurs—whether we like it or not—when substantive problems that two or more institutions address are linked in biogeophysical or socioeconomic terms. The politics of institutional design and management, in contrast, comes into play when actors forge links between issues and institutions intentionally in the interests of pursuing individual or collective goals (Young et al. 1999). Combining the two dichotomies produces a 2 × 2 table that provides a conceptual map of institutional interplay (table 1.1).[2]

Identifying vertical interplay requires an explicit demarcation of boundaries separating different levels of social organization. It is common, in this connection, to start by distinguishing among micro-, meso-, and macroscale systems or, in other words, local, national, and international levels of social organization. But this classification is far from precise, much less objectively correct. National systems range from microstates

Table 1.1
Types of institutional interplay

	Functional interdependencies	Politics of design and management
Horizontal	UNFCCC, ozone regimes	Joint funding mechanisms (e.g., GEF)
Vertical	CBD, national forest regimes	CLRTAP, national air pollution regimes

Note:
UNFCCC: United Nations Framework Convention on Climate Change
GEF: Global Environment Facility
CBD: Convention on Biological Diversity
CLRTAP: Convention on Long-Range Transboundary Air Pollution

(e.g., Luxembourg, Nauru) to continental states (e.g., Russian Federation, United States), and some local jurisdictions (e.g., North Slope Borough in Alaska) cover larger areas than many nation states. Nor is anything sacred about the tripartite division of levels of social organization into local, national, and international systems. Although social scientists have used this simple scheme for many purposes, interest is growing in regional arrangements operating above the local level but below the national level, as well as in regional arrangements operating above the national level but below the global level. Analyses of vertical interplay do not presuppose anything about appropriate or preferred ways to differentiate among levels of social organization. They simply direct attention to cross-scale interactions, however the relevant levels of social organization are defined.

Whereas vertical interplay turns on distinctions among levels of social organization, horizontal interplay emphasizes the importance of differentiating between or among institutions operating at the same level. Because institutional arrangements often run into each other at the margins, it is not always easy to determine where one institution ends and another begins, and this can become increasingly complex as rules evolve with the passage of time. As the global trade regime has taken on environmental provisions and a variety of environmental regimes (e.g., arrangements dealing with ozone depletion, hazardous wastes, and endangered species) have taken on provisions dealing with trade, for instance, the boundaries between and among these arrangements have begun to blur. It follows that separating distinct regimes operating at the same level can be a tricky business. Yet no one doubts that useful distinctions are to be drawn in this realm, and it is the separation between functionally or spatially distinct arrangements that opens up the prospect of horizontal interplay.

Long familiar in domestic settings, functional interdependencies are rapidly becoming an important concern at the international level as well. International regimes dealing with ozone depletion and climate change, for instance, are linked functionally because chlorofluorocarbons (CFCs), which are the central concern of the ozone regime, are also potent greenhouse gases and because a number of the chemicals that seem attractive as substitutes for CFCs are also greenhouse gases (Oberthür 1999). Regimes dealing with regulation of marine pollution and with protection of stocks of fish and marine mammals are functionally linked because the success or failure of efforts to control pollution can be expected to have significant consequences for the well-being of marine ecosystems and stocks of fish and other organisms they support. For that matter, regimes that regulate fishing and those designed to protect marine mammals are functionally linked as a consequence of the fact that whales, seals, and other marine mammals are dependent on fish as a food source and often suffocate when they become entangled in fishing gear.

A number of distinct motives can lead actors to engage in deliberate attempts to link institutions at the stages of design and management (Young 1996). Such initiatives sometimes arise from a desire to improve the performance of individual regimes. Efforts to nest local or regional arrangements (e.g., various regional seas regimes) into larger or more comprehensive arrangements (e.g., overall law of the sea), for instance, typically rest on a belief that the effectiveness of the smaller-scale arrangements will be enhanced by integrating them into larger systems. In other cases, political linkages arise from efforts to improve efficiency by integrating the supply of services necessary to operate two or more institutional arrangements. Funding mechanisms and dispute-settlement procedures are familiar cases in point. The Global Environment Facility, for instance, provides funding for both the climate regime and the regime dealing with protection of biological diversity (Sand 1999). At the same time, political linkages are common when actors seek to use a second arena to gain advantages with regard to pursuit of interests that are blocked in a primary arena. Recent efforts on the part of those favoring resumption of (limited) commercial whaling to promote their goal through the regime for trade in endangered species constitute an example.

The occurrence of functional interdependencies often suffices to trigger the emergence of political interplay. Faced with severe side effects or mutual interference, actors may find compelling reasons to address these issues in the context of institutional design. The same can be said regarding efforts to create vertical hierarchies of institutions operating in a single issue area. But the existence of such linkages is not a necessary condition for the emergence of linkage politics. A particularly interesting illustration arises when actors devise packages or clusters of institutional arrangements largely for strategic purposes rather than as instruments for coming to terms with functional interdependencies. Matters pertaining to navigation, fisheries, offshore oil and gas development, deep seabed mining, pollution control, and scientific research are often dealt with by distinct systems of rules that work perfectly well in their own domains. But in specific cases, actors may find it expedient to combine these concerns into comprehensive sea-use regimes to negotiate package deals that are acceptable to all stakeholders with legitimate claims to use the resources (Sebenius 1983). As the comprehensive law of the sea set forth in the 1982 convention makes clear, complex packages are often difficult to negotiate, much less to implement once their provisions have been ratified (Friedheim 1993). Over time, however, they may prove helpful in coming to terms with common problems of institutional interplay that arise in areas in which various individual arrangements have been created with little concern about their implications for related arrangements.

The Problem of Scale
Scale has to do with the levels at which phenomena occur in the dimensions of space and time. Much work on regimes dealing with CPRs, for instance, is based on the study of small-scale, typically local arrangements devised to deal with human uses of natural resources such as stocks of fish, water, trees, or grazing lands. At the same time, many observers have noted the fact that some global systems, such as the electromagnetic spectrum or Earth's climate system, also exhibit the defining features of CPRs (Sandler 1997). It is natural, under the circumstances, to ask whether propositions derived from the study of small-scale systems apply to global CPRs as well and vice versa (Young 1994b). Note that this issue differs from the central concern of the problem of interplay. Institutional arrangements often interact with one another across levels of social organization, giving rise to more or less complex forms of vertical interplay. But the problem of scale is not a matter of interactions or linkages among distinct institutions. Rather, it centers on the extent to which the dynamics of systems that differ from each other in terms of spatial or temporal scales are nonetheless sufficiently similar so that we can scale up and down in seeking to understand how they work.

In the analysis of human systems, spatial scale is a familiar concept. In fact, disciplines such as political science employ clear distinctions based on scale (e.g., local, national, and international politics) as a means of differentiating important subfields. The implicit assumption is that scale matters in the sense that important differences exist between local and national polities or between national polities and politics at the international level. In the case of political science, this distinction turns, for the most part, on the character of political institutions prevailing at different levels of social organization. Thus, the presence or absence of a state is treated as a matter of such fundamental importance that it is unlikely that cross-scale comparisons between local and national or between national and international systems will prove insightful. Yet the study of governance systems created to deal with environmental problems raises interesting questions about this presumption (Young 1999b). To the extent that design principles derived from a study of small-scale CPRs are also helpful in thinking about avoiding tragedies of the commons at the global level (Ostrom et al. 1999), for instance, spatial scale will emerge as an interesting focus of analysis for those concerned with the institutional dimensions of environmental change.

Temporal scale is a different matter. One of the central puzzles for those seeking to understand the behavior of Earth's climate system, for instance, involves the relative importance of interannual cycles (e.g., El Niño events), decadal cycles (e.g., variations in solar radiation), and millennial cycles (e.g., advances and retreats of glaciers and ice sheets) that operate simultaneously—albeit at different temporal scales—but appear to involve mechanisms that are quite different. With regard to human systems, a number of observers have attempted to document business cycles; some have sought to identify electoral cycles or issue-attention cycles, and a few have suggested that it is possible to discern "long cycles" relating to the rise and fall of great powers at the international level or even the rise and fall of human civilizations as portrayed by writers such as Spengler and Toynbee (Downs 1972; Goldstein 1986; Kennedy 1987). Such efforts constitute a minor stream of analysis, and obvious differences among the cycles in question are so great that few have thought to pose specific questions relating to cross-scale comparisons. Nonetheless, temporal scale is an interesting topic for those concerned with the institutional dimensions of environmental change. To the extent that considerations of temporal scale are relevant to biogeophysical systems, institutional arrangements have to include procedures for tracking processes occurring at different scales (e.g., interannual fluctuations in temperature due to natural variability versus decadal changes associated with the greenhouse effect). And should it turn out that temporal scale is relevant to human systems as well, this may have additional implications for creation and implementation of institutions. It is perfectly possible, for example, that formation of institutions occurs in waves, with periods of innovation (e.g., 1970s, 1990s) alternating with periods of consolidation (e.g., the 1980s, 2000s).

The Road Ahead

Many individual studies shed light on the roles that institutions play both as drivers of environmental change and in responses that humans make to environmental problems. Clearly, there is ample scope for the flow of these individual studies to continue. At the same time, this subject lends itself to a large-scale, programmatic effort on the part of researchers trained in a variety of fields of study but willing to engage in a concerted effort to make progress toward answering questions of causality, performance, and design outlined in this chapter. The key to success is to strike a proper balance between coordinating activities of numerous researchers and leaving individual researchers ample scope for imagination and creativity. Individual projects that are not based on compatible definitions of key concepts and common formulations of central questions run the risk of yielding results that are not comparable, even when they seem on the surface to address the same topic (e.g., the role of institutions as causal forces). Yet any attempt to impose too much order on the scientific endeavors of individuals not only suppresses innovation, it is virtually certain to break down in dissension. Familiar to those working in many of the natural sciences, this challenge is relatively new in most social sciences. But success in meeting this challenge is likely to make all the difference in determining our ability to produce both basic knowledge and policy-relevant insights pertaining to the institutional dimensions of environmental change.

2

Collective-Action Models versus
Social-Practice Models

How can institutions, which are not actors in their own right, guide the behavior of those whose actions give rise to environmental problems and in so doing play significant roles in solving (or, for that matter, causing) problems ranging from the depletion of living resources (e.g., fish and forests) to the disruption of large-scale ecosystems (e.g., Earth's climate system)? To phrase the question in more generic terms, what are the mechanisms through which social institutions or governance systems affect the course of interactive decision making among two or more human actors in a variety of settings? To pose these questions is to launch an enquiry into the role of institutions as causal forces in human affairs. Practitioners of all the social sciences have devoted increased attention to these questions in recent years and produced answers that seem convincing to their own colleagues. Yet the overall result has been a proliferation of disparate perspectives rather than an emerging consensus concerning the causal role of institutions.

Those whose interests in these matters stem from a desire to understand and ultimately to solve large-scale environmental problems have developed two principal clusters or families of models intended to illuminate the causal role of social institutions and, in the process, to account for variations in the effectiveness of specific environmental regimes. One cluster—I call them collective-action models—encompasses constructs that draw on the intellectual capital of economics and public choice and treat actors as decision makers basing their choices on utilitarian calculations (Rutherford 1994). The other cluster—I call them social-practice models—includes constructs that stem from anthropology and sociology and emphasize the roles of culture, norms, and habits as sources of behavior

(Powell and DiMaggio 1991). As is true of all analytic as well as genetic groupings, members of these families differ significantly. Some differences are rooted in analytic distinctions; others merely typify the interests of those who have created particular models. The social-practice cluster, in particular, is an extended family featuring substantial variation among its individual models. Yet members of each family have enough in common to make it helpful to group them into these two broad clusters.

In this chapter I highlight essential differences between these two ways of thinking about the role of social institutions as causal forces, explore implications of these differences for our understanding of the effectiveness of environmental regimes, and consider prospects for evaluating explanations derived from both models in the interest of enhancing our ability to solve environmental problems. Overall, the analysis yields mixed results. Each cluster captures significant elements of reality; neither is sufficient by itself to provide an adequate basis for understanding the institutional dimensions of environmental change. It follows that the two families are complementary. At this stage, we need them both; nothing is to be gained from approaching this issue in either-or terms. Wherever possible, I turn to large-scale environmental concerns (e.g., ozone depletion, climate change, loss of biological diversity, disruption of large marine ecosystems, together with international regimes created to address them) as illustrations of the analysis to follow (Victor and Salt 1994; Mitchell 1995; Parson and Greene 1995; Raustiala and Victor 1996). But the basic argument applies wherever institutions emerge as significant drivers in causing or confronting environmental change.

Differentiating the Models

Proponents of the new institutionalism have articulated many overlapping although by no means identical definitions of institutions. For purposes of this discussion, however, it suffices to say that institutions are sets of rules, decision-making procedures, and programs that give rise to recognized practices, assign roles to participants in these practices, and govern interactions among occupants of specific roles (Young 1994a). But what is it that makes institutions and, more specifically, regimes dealing with large-scale environmental concerns more or less effective in the sense of exerting influence over outcomes arising from interactions among their members as well as those operating under the auspices of their members?

One way to approach this question is to think of regimes as devices created by actors seeking to avoid or alleviate collective-action problems (Hardin 1982). It is well known that interactive decision making among actors striving to maximize their own welfare often produces outcomes that are suboptimal for all concerned, or less desirable for all participants than one or more feasible alternatives. Although the situation known as prisoner's dilemma is commonly treated as the paradigmatic collective-action problem, numerous circumstances generate incentives to act in ways that seem rational from an individualistic perspective but yield collective outcomes that are unattractive to all (Schelling 1978). The best-known environmental exemplar has become familiar to analysts and practitioners alike as the tragedy of the commons (Hardin 1968; Hardin and Baden 1977; Baden and Noonan 1998). Regimes on this account are arrangements that those experiencing or expecting to experience collective-action problems create in the interest of avoiding joint losses or reaping joint gains. They are successful or effective to the extent that their operation lowers the probability of uncooperative behavior of the type that leads to losses of welfare for all parties concerned.

Another way of thinking about the role of regimes as causal forces is to treat them as arrangements giving rise to social practices that shape the identities of participating actors, generate common discourses in terms of which to address environmental problems, and draw participants into routinized activities that do not involve utilitarian calculations on a day-to-day basis (Powell and DiMaggio 1991; Wendt 1999). On this account, rules and decision-making procedures affect behavior in part by performing constitutive functions (Kratochwil 1989; Onuf 1989). But even when actors have well-defined identities that are independent of the provisions of particular regimes, creation of a new regime can shape the ways participating actors think of their roles and the willingness of others to accept the appropriateness of the resultant role behavior. The law of the sea crystallized in the 1982 U.N. Convention on the Law of the Sea (UNCLOS), for instance, creates exclusive economic zones (EEZs) covering about 11 percent of the world ocean and accords extensive, although not unlimited, jurisdiction over these zones to the adjacent coastal states. States that have accepted this institutional arrangement do not engage in calculations on a case-by-case basis regarding benefits and costs of acknowledging the existence of specific EEZs. For the most part, they simply accept the jurisdiction of coastal states in these zones and consider their options on the basis of this assumption.

An effort to unpack these general formulations reveals that collective-action and social-practice models diverge in terms of the assumptions they make about several different matters. For ease of exposition, I divide these assumptions into three broad categories: assumptions about the identity and character of relevant actors, about sources of actor behavior, and about the social environment and extent to which actors are affected by the operation of social constraints.

Actor Identity

Collective-action models focus on the behavior of regime members treated for the most part as unitary actors whose identities, often described in terms of role premises or utility functions, predate and are largely unaffected by participation in specific institutional arrangements. In the case of international regimes, members are signatories to constitutive agreements, including conventions, treaties, and formal declarations, that bring institutional arrangements into existence. Ordinarily, these actors are states, as in the cases of the signatories to UNCLOS, the 1992 U.N. Framework Convention on Climate Change (UNFCCC), and the 1992 Convention on Biological Diversity. Unitary actors are behaving units that have integrated utility functions and that make choices among available options in such a way as to promote their own welfare. States, on this account, exist over long periods of time; they are likely to be members of numerous regimes dealing with environmental concerns as well as a variety of other matters. Entry into a particular regime may affect the incentives of these actors as they assess the benefits and costs of available options in specific situations. But membership will not have significant effects on the overall identity or general interests of states whose existence not only predates formation of specific regimes but also rests on a variety of considerations that extend well beyond the purview of regimes dealing with specific problems. Thus members make utilitarian calculations regarding the benefits and costs of complying with rules of specific regimes and of living up to commitments they made in joining the regimes; they do not approach these matters as obligations or duties to be met as a matter of course.

Social-practice models differ with regard to all these assumptions about the identity and nature of the actors. Whereas states may be members of most international regimes in formal terms, actors whose behavior gives rise to environmental problems and whose responses are critical to solving them typically include corporations, nongovernmental organizations, and even individuals. Thus, all consumers of energy produced through the burning of fossil fuels, from multinational corporations down to individual home owners, generate emissions of greenhouse gases that enter Earth's atmosphere. A key concern in thinking about environmental regimes, then, is to understand how states as formal international members are able to use their membership to influence the behavior of nonstate actors (Victor, Raustiala, and Skolnikoff 1998). Nor is it helpful to regard states themselves as unitary actors pure and simple. Although it may do no harm to conceive of them in this way for some purposes, states are complex actors whose constituent elements have different and sometimes conflicting interests with regard to individual regimes. It will come as no surprise that public agencies whose mandates cover environmental protection will react differently to terms of an agreement such as the UNFCCC than agencies whose objectives are framed in terms of promoting the interests of commerce and industry. The familiar metaphor of the two-level game only scratches the surface of this phenomenon (Putnam 1988). Even more fundamental is the assumption common to social-practice models that membership in regimes often has constitutive effects. The issue has been debated extensively in recent years under the rubric of the agent-structure problem (Wendt 1987; Dessler 1989). But the key point for this discussion concerns the extent to which actors express their interests and even conceive of their identities in terms of regime membership.[1] An actor that considers participation in a regime an important part of its role will not organize its thinking in terms of questions about compliance and living up to commitments; it will engage in behavior deemed appropriate to the role it occupies as a matter of course (Wendt 1999).

Sources of Behavior

Collective-action models rest on utilitarian premises in the sense that they regard actors as decision-making units approaching choices in cost:benefit terms and seeking in each case to select the option that will best promote their own welfare and, more specifically, maximize the net benefits to be derived from the choice. In this general framework, many variations are possible. Some models assume that options available to decision makers are produced exogenously and not subject to change; others allow the actors themselves to exercise some influence over both framing problems and identifying options. Because welfare is a highly abstract concept that is notoriously difficult to operationalize, collective-action models commonly resort to some surrogate, such as income or wealth defined in monetary terms. Differences also relate to the extent to which actors engaged in interactive decision making are treated as maximizers of absolute gains or of relative gains (Baldwin 1993). Are states involved in efforts to protect Earth's climate system or to maintain biological diversity more concerned with their standing relative to that of others, or with the condition of the environment in absolute terms? There are no simple, much less correct, answers to questions of this sort. But in every case, collective-action models assume that actors are driven by the logic of consequences in the sense that their choices are based on efforts to assess benefits and costs associated with individual options.

Here, too, social-practice models rest on different premises. The differences are captured in part in the idea of the logic of appropriateness as opposed to the logic of consequences (March and Olsen 1998). Thus, social-practice models often assume that actors comply with rules or live up to commitments because they are authoritative or legitimate or, to put it another way, because such behavior is deemed normatively correct or proper (Franck 1990). But there is more to the distinction between the two families of models than this. Many social-practice models incorporate an alternative source of behavior by assuming that actors commonly adhere to the rules of regimes as a matter of habit or because such behavior is taken for granted as a result of socialization or routinization. Even more fundamental in these terms are consequences of discourses and role definitions. States that respect each other's EEZs generally do so because they tend to comply with international law as a matter of course and because they accept the proposition that coastal states possess jurisdiction over adjacent marine areas, so that the authority to govern specific activities occurring within the EEZ is an accepted component of the role of coastal state. Yet behavior that is routinized need not become fixed or unchanging. Sociological perspectives regularly include the prospect of social learning (Haas and Haas 1995; Social Learning Group 2001) with regard to both matters of fact (e.g., impact of greenhouse gas emissions on Earth's climate system) and normative concerns (e.g., the extent to which individual actors have an obligation to refrain from disturbing global commons). Social-practice models encompass a wider range of perspectives on the sources or roots of actor behavior than collective-action models. Whereas collective-action models all share the tenets of utilitarianism, it is fair to say that social-practice models constitute a more extended family that includes several distinct branches. But what joins social-practice models as a single family and sets them apart from members of the collective-action family is the assumption that behavior is driven by forces that do not feature efforts to calculate benefits and costs and to make choices in such a way as to maximize net benefits.

Social Constraints

Collective-action models assume either that specific choices stand alone in the sense that they are not embedded in a larger social environment, or that the effects of the social environment can be endogenized in calculations actors make regarding benefits and costs associated with specific available options. What does this mean in practice? Consider a game-theoretic formulation such as prisoner's dilemma. All the information necessary to determine the outcome of the interactive decision-making process is included in the matrix representing the game in normal form (Luce and Raiffa 1957). The implication is that individual participants do not have access to information about one another, about the history of their relationship, or about the larger social setting in which they operate except insofar as such information is embedded in the matrix. Nor is such information considered relevant in determining the content of collective outcomes. To be sure, this is an extreme case. But it illustrates what is meant by the observation that relationships described in collective-action models are self-contained.

In social-practice models, in contrast, context not only matters but it often becomes a source of external or exogenous constraints on the behavior of those engaged in interactive decision making regarding a particular regime. Thus, regimes dealing with control of ozone-depleting substances and protection of biological diversity are embedded in the larger setting of international society whose members are sovereign states accustomed to exercising unrestricted authority over their own internal affairs (Ruggie 1983; Hurrell 1993). Any effort to devise rules that allow outsiders to intervene in the affairs of member states in the name of protecting stratospheric ozone or biodiversity is likely to arouse vigorous opposition justified in terms of rules and norms pertaining to sovereignty in a society that lacks a higher authority (Hurrell 1992; Conca 1995). Similarly, changes occurring at this broader level may well prove significant for the operation of specific regimes. If, as some observers maintain, sovereignty is presently undergoing major changes that enhance the acceptability of efforts to impose international standards or norms on individual members of international society, this will have far-reaching consequences for efforts to devise effective regimes aimed at solving problems such as loss of biological diversity (Lyons and Mastanduno 1995; Litfin 1998). The same holds true for other features of the broader social environment, such as the status of nonstate actors (Cutler, Haufler, and Porter 1999) or the acceptance of general rules pertaining to the interpretation and implementation of international conventions and treaties. But in each case, the basic point is the same. The performance of specific regimes dealing with more or less well-defined problems is sensitive to a variety of constraints emanating from the broader social environment in which they operate.

A major virtue of collective-action models is that they are relatively compact and parsimonious, which makes them more tractable and more conducive to formalization than social-practice models. Whereas social-practice models point to a number of types of actors and several distinct sources of behavior, collective-action models direct attention to easily identifiable and relatively small sets of unitary actors whose behavior is based on familiar precepts of utilitarianism. This is surely an advantage, especially for those interested in the development of formal theories regarding roles that institutions play as causal forces. But for those whose primary goal is to understand how international regimes can solve environmental problems and why some regimes prove more successful than others, a substantial price must be paid for this advantage. In part, this is a matter of operationalization. It is difficult, under real-world conditions, to provide empirical content for utility functions of states in a manner that does not involve arbitrary procedures. In part, the problem stems from the importance of factors that are typically omitted from collective-action models. With regard to problems such as climate change and loss of biological diversity, for instance, a critical issue centers on the ability and willingness of states to translate international commitments into domestic rules and regulations capable of redirecting the behavior of those—including corporate executives, operators of municipal power plants, and individual homeowners or car owners—whose actions give rise to emissions of greenhouse gases or destruction of habitat vital to different species. To the extent that such factors turn out to be major determinants of the effectiveness of international regimes, efforts to explain levels of effectiveness that rely solely on collective-action models will fail. Under the circumstances, analysts of international regimes have good reasons to pay increased attention to social-practice models, even while continuing to explore the uses of the more elegant collective-action models.

Exploring the Implications

What are the implications of these differences in the theoretical foundations of the two families of models for efforts to understand the role of institutions as causal forces in international society? More specifically, how does the choice of models affect our expectations regarding the ability of international regimes to solve environmental problems? I make no effort in this section to present a comprehensive account of differences in predictions flowing from specific models of interactive decision making. Rather, I examine a few key issues that have proved contentious among students of international regimes and show how the choice of theoretical assumptions affects our thinking about these matters. Specifically, I focus on four main themes: compliance with and fulfillment of commitments, consequences of different policy instruments, behavioral consistency, and durability of regimes.

Compliance

Analysts who work with collective-action models generally regard compliance with regime rules and fulfillment of international commitments as problematic. Whereas they naturally agree that compliance is the rational choice for utility maximizers in some situations, they are likely to see the underdeveloped character of enforcement mechanisms in international society as a critical problem that will limit—if not undermine—the effectiveness of environmental regimes. And the deeper or more extensive the cooperation, the more serious this problem is likely to become (Downs, Rocke, and Barsoom 1996; Barrett 1999). Those who operate from a social-practice perspective, in contrast, typically maintain that enforcement is not the key to compliance and argue that there is nothing unusual or troubling about international society with regard to achievement of compliance with regime rules or fulfillment of commitments (Chayes and Chayes 1995).

To be concrete, members of the first group can be expected to maintain that success in dealing with climate change requires not only adoption of precise targets and timetables but also development of a meaningful system of sanctions to ensure that subjects live up to these commitments. Those who think in social-practice terms are more likely to focus on the question of whether the climate change regime promotes emergence of a discourse that comes to dominate the way in which parties think about the issue and draws actors into a set of procedures for dealing with emissions of greenhouse gases that are increasingly taken for granted. The consequences of the regime, on this account, will flow from routinized behavior rather than from explicit decisions about compliance. What are the sources of this difference of views regarding compliance, and what would be required to test the relative merits of these perspectives (Mitchell 1996)?

Utilitarian actors will comply with rules and fulfill commitments if, and only if, they are convinced that the present value of compliance exceeds the value of noncompliance (Young 1979). They will routinely discount the costs of noncompliance by taking into account the probability that violations will go undetected and that sanctions will be modest even when they are detected. Given the well-known attractions of free riding or, in other words, the prospect of reaping benefits of cooperation even while failing to comply oneself (Olson 1965), it is to be expected that actors will often discount the benefits associated with compliance as well. None of this means that problems of compliance will always be acute in such settings. In the case of rules developed to solve coordination problems where there is no incentive to cheat, even utility maximizers will comply as a matter of course (Stein 1982).

Under the circumstances, however, it will come as no surprise that students of international regimes search vigorously for mechanisms that will increase the benefits of compliance or raise the costs of noncompliance (Oye 1986). Efforts to enhance transparency can make it difficult for actors to violate rules clandestinely. Measures aimed at lengthening the shadow of the future can raise the costs of noncompliance, especially for actors using relatively low discount rates in calculating the present value of future costs and benefits (Axelrod 1984). Linking institutional arrangements together can raise the price of noncompliance by increasing the probability that the consequences of violations in one issue area will spill over in such a way as to degrade cooperation in other areas of importance to individual actors. But useful as such devices are in specific situations, it is easy to see why those who approach compliance in these terms often conclude that the availability of credible sanctions is a key determinant of the effectiveness of international regimes. It is worth noting that sanctions need not take the form of penalties administered by agents authorized or licensed by international society to play this role. The threat of reprisals or retaliatory actions carried out by other regime members, or even by nonstate actors, may be sufficient to deter noncompliant behavior in some situations. But this does nothing to reduce the emphasis on the role of enforcement among those who think in these utilitarian terms.

Several factors account for the comparatively relaxed attitude toward this problem characteristic of those who think in social-practice terms. To the extent that regimes influence the interests and even identities of their members, the problem of persuading actors to comply with their rules and procedural requirements does not arise. Just as it makes no sense to think of a chess player refusing to comply with the rules of chess, there is little reason to worry about cheating or noncompliance with rules of regimes that have powerful constitutive effects. Short of this, however, social practices regularly produce compliant behavior through mechanisms featuring the operation of feelings of propriety on the one hand and the development of habits or routinized behavior on the other. To the extent that rules are accepted as legitimate or authoritative, those whose actions are governed by the logic of appropriateness rather than the logic of consequences will comply with them without making an effort to calculate whether the benefits of doing so outweigh the costs. Equally important is the impact of socialization or the processes through which compliance becomes habitual or routine in the sense that subjects engage in required behavior or refrain from proscribed behavior without thinking about the matter on a case-by-case basis (Hart 1961). In effect, this line of thinking suggests that socialization plays much the same role in international society as it does in other social settings. Calculated decisions constitute the exception rather than the rule. Under ordinary circumstances, actors comply as a matter of course or develop what students of public policy sometimes call standard operating procedures that guide their actions in the absence of conscious decision making (Allison 1971).

Policy Instruments

A vigorous debate has arisen between those who favor policy instruments based on command-and-control regulations and those who prefer what are known as incentive mechanisms. As applied to international environmental regimes, incentive mechanisms take the form of agreements on overall targets (e.g., a percentage reduction in emissions of greenhouse gases, commitment to preserve habitat of critical importance to the protection of biodiversity) that allow subjects to make their own choices about how to fulfill these commitments. Command-and-control regulations feature agreements that require subjects to take specified actions (e.g., installing segregated ballast tanks in new oil tankers, incorporating smokestack scrubbers into the design of new power plants). Broadly speaking, those who work with collective-action models take the view that incentive mechanisms are preferable to command-and-control regulations not only in terms of effectiveness (getting the job done) but also in terms of efficiency (achieving desired results at the lowest cost). Those who favor social-practice models are more favorably inclined toward command-and-control regulations. What are the sources of this difference, and what can we say about the testability of the two views?

For those who think in utilitarian terms, everything comes down in the final analysis to incentives. But the specific case for incentive mechanisms as a source of effectiveness in international environmental regimes rests on a combination of arguments focused at the individual level and propositions framed in terms of social welfare (Opschoor and Turner 1994; Rose 2002; Tietenberg 2002). Incentive mechanisms allow individual subjects to make their own choices about how to respond to overall goals and targets and, in the process, to minimize the costs of complying with regime rules and commitments. In effect, encouraging actors to pursue efficiency in responding to requirements and prohibitions bolsters effectiveness by lowering costs to individual actors of adjusting their behavior in such a way as to conform to prescriptive provisions of institutional arrangements. Handled properly, incentive mechanisms can also become a source of revenue for governments of states that are members of a regime.[2] Of course, such revenue may simply flow into the general fund and have no effect on the performance of environmental regimes. Nonetheless, this prospect does suggest that incentive mechanisms can provide material resources required to assist subjects that are weaker—domestically as well as internationally—to live up to regime rules and commitments. Among other things, this can ease the burden on wealthier regime members arising from the adoption of effectiveness-enhancing measures, such as the compensation fund (officially, the Montreal Protocol Multilateral Fund) of the ozone regime and the arrangements dealing with technology transfer of the climate change regime.

Those who work with social-practice models, on the other hand, find the use of incentive mechanisms troubling, and for several reasons. Use of such mechanisms as a standard procedure may have the effect of making commodities of compliance and fulfillment of commitments. Encouraging subjects to respond to obligations in utilitarian terms can legitimize behavior on the part of individuals that falls well short of overall requirements. In extreme cases, it may even lead actors to conclude that behavior on their part that causes depletion of living resources or pollution of ecosystems is acceptable, as long as they are willing to pay for it through such procedures as joint implementation or payment of fees or fines that can subsidize environmentally benign actions on the part of others. The result is a condition in which the role of legitimacy or authoritativeness as a force conducive to compliance or fulfillment of commitments is diminished, perhaps seriously. What is more, relying on incentive mechanisms will encourage subjects to make careful and repeated calculations about the relative merits of alternative responses to institutional prescriptions, a condition that raises questions about the role of habitual or routine behavior in living up to the terms of commitments. On this account, a certain amount of inefficiency regarding the selection of means to be used in achieving goals is a small price to pay for a high level of conformance on the part of individual subjects (Mitchell 1994).

Behavioral Consistency

Another issue on which differences are sharp in expectations flowing from theoretical models involves behavioral consistency. Those who adopt a collective-action perspective, especially its unitary-actor variant, generally expect actors to make careful calculations about participation in individual regimes at an early stage and then to exhibit a high degree of consistency with respect to actions taken on the basis of conclusions flowing from these calculations. To them, the behavior of the United States in refusing initially to sign the Convention on Biological Diversity (CBD), then reversing itself on this matter, and finally refusing to ratify the convention is anomalous. Much the same is true of the behavior of Australia and France in agreeing to the Antarctic minerals convention only to turn against the agreement and begin to campaign for its replacement within months after accepting the final text (Stokke and Vidas 1996; Joyner 1998). From a social-practice perspective, nothing is surprising about behavior of this sort. In fact, those who think in these terms are likely to look on behavioral consistency on the part of regime members as a puzzle or a proper target of analysis. What accounts for this contrast, and which view comes closer to matching reality in the world of international environmental regimes?

The collective-action perspective on behavioral consistency flows directly from the assumption that regime members are unitary actors whose identities and basic interests are not affected significantly by participation in regimes dealing with specific problems. Of course, actors of this sort can experience shifts in calculations they make about participation in individual regimes. New information about the effects of climate change may lead unitary actors to become more (or less) interested in implementing the provisions of measures such as those set forth in the 1997 Kyoto Protocol to the U.N. Framework Convention on Climate Change. Unitary actors may even experience changes in the evaluative criteria they use in judging the significance of factual information. There is no reason to assume that they are incapable of learning when it comes to the treatment of environmental problems. Even so, it is reasonable to expect a relatively high level of behavioral consistency on the part of those who have generally stable utility functions and are not plagued with internal dissonance. From this perspective, the rather consistent role of the United States as a laggard with regard to development of international measures to cope with climate change seems perfectly understandable.

The key to the social-practice perspective on this matter lies in the fact that it pays particular attention to actors treated as collective entities whose behavior with regard to international issues is the product of (often complex) internal interactions among a variety of interest groups or stakeholders.[3] The behavior of the United States in the case of the CBD is easy enough to understand, for instance, in terms of the transition from the Bush administration to the Clinton administration, followed by a sharp shift in the composition of Congress that undermined prospects for ratification of this convention. And this is a comparatively simple case of the internal dynamics of collective actors that can, and often do, make behavioral consistency with respect to participation in environmental regimes the exception rather than the norm. Quite apart from these political considerations, institutions regularly take on lives of their own with the result that rules and the discourses that underpin them change substantially over time. It is no surprise that individual actors vary in their responses to such developments. Some take the lead in encouraging change; others become strict constructionists seeking to defend the regime in its original form. In many instances, however, developments of this kind trigger vigorous debates within individual regime members, a process that can promote various forms of behavioral inconsistency as one or another faction gains the upper hand.

Durability

Students of collective action typically regard international regimes as fragile arrangements that are likely to fall by the wayside whenever one or a few key members lose interest or become disillusioned with their performance. In extreme cases, they are likely to treat regimes as epiphenomena (Strange 1983) and to chide those who are optimistic about the role of regimes in solving international problems for falling prey to "the false promise of institutions" (Mearsheimer 1994–1995). Here, again, the contrast between this view and the social-practice perspective is striking. Those who think in terms of social practices typically regard institutions as sticky or persistent; many arrangements are highly resistant to pressures for change, even when such pressures emanate from actors that are undoubtedly influential members of the relevant groups. Institutions often acquire lives of their own that allow them to exercise influence over the behavior of members and other subjects long after the circumstances that led to their creation have disappeared. To take a concrete example, many students of collective action regard the climate change regime as a delicate construct that may well collapse or fade into obscurity if the Kyoto Protocol does not enter into force or if the United States does not become a more cooperative partner in this enterprise. Those who start from a social-practice perspective point to the evolution in the regime that has occurred since the signing of the UNFCCC in 1992 and see evidence that participants are already becoming enmeshed in a complex set of activities that they cannot easily ignore or disown. What accounts for this divergence, and what steps might be taken to test the merits of the two perspectives?

One way to characterize the collective-action perspective on this issue is to say that it presents a picture in which regimes are lightly institutionalized. Specific regimes are surface manifestations of bargains struck among key members of the group of actors concerned with the issue at hand (Strange 1983). As a result, changes in the distribution of bargaining strength among major participants or alterations in the positions of one or more key players are likely to erode the political foundations of arrangements created to deal with major problems. In recent years, the volatile and often obstructionist policies of the United States have presented a particularly severe challenge to the development of effective environmental regimes. Thus, the refusal of the United States to ratify the CBD, together with the more recently negotiated protocol on biosafety, raises profound questions about the viability of the biodiversity regime. Somewhat similar comments can be applied to the unwillingness of the United States to ratify the Kyoto Protocol on climate change. Regimes have little capacity to control their own destinies; their fortunes rise or fall in response to fluctuations in both the capacity and the political will of their members or other actors who are influential in the relevant area. International regimes are typically short-term arrangements whose contributions to solving environmental problems remain quite limited.

For its part, the social-practice perspective points to the roles of shared discourses, socialization, and institutional cultures to account for the stickiness of institutional arrangements. The point is not that institutions become rigid or inflexible, although certainly in some cases regimes fail to adapt successfully to important changes affecting the issues they address. As many observers point out, rules often evolve in ways that are hard to connect to formal provisions of constitutive agreements even though subjects may understand them perfectly well (Ostrom 1990). But the essential point here is that participation in a regime is likely to influence the way members think about their roles and lead to an institutional culture that guides the actions of relevant actors in the absence of specific calculations dealing with benefits and costs of participation. Those who adopt this perspective are impressed by the considerable evolution that occurred in the climate regime during the 1990s. In less than a decade this regime became the center of a remarkable growth in international activities linked to the regime informally in some instances (e.g., assessments of the Intergovernmental Panel on Climate Change) and more formally in others. In this case, failure of the United States to ratify the Kyoto Protocol does not loom as a fatal flaw. Despite its reluctance to accept formal commitments regarding emissions of greenhouse gases, the United States has contributed greatly to the growth of scientific knowledge relating to climate change and may find itself enmeshed in the evolving regime in ways that are difficult to ignore or escape.

It seems accurate to conclude that social-practice models accord institutions a larger role than collective-action models as causal forces in shaping the course of events in situations involving interactive decision making. This difference arises from the fact that social-practice models point to mechanisms that have deeper behavioral roots than utilitarian calculations that constitute the core of collective-action models. Of course, it would be a mistake to infer that social-practice mechanisms will lead to greater success in solving international environmental problems than collective-action mechanisms. Sticky institutional arrangements often become part of the problem in dealing with such issues rather than part of the solution (Young et al. 1999). Deeply entrenched international rules that are routinely incorporated into regimes dealing with specific problems, such as the prohibition against interference in the domestic affairs of member states and the proviso that states cannot be forced to accept commitments against their will, are obviously serious constraints when it comes to designing regimes dealing with matters such as climate change. This is true especially of biodiversity, where a major source of the problem involves actions occurring within member states that their governments are unable or unwilling to regulate.

Evaluating Divergent Expectations

It is difficult to devise tests that will yield unambiguous conclusions about the relative merits of these families of models and that unbiased observers will accept as decisive. A number of factors contribute to this difficulty. Significant questions arise about what it means to test families of models in contrast to individual models. It is hard to distill competing hypotheses from specific models that are formulated in sufficiently precise and operational terms to allow for straightforward testing. Debates are continuing about the boundaries of the universe of cases—the set of international environmental regimes—to which these hypotheses should apply, and considerable heterogeneity exists among cases that most analysts would agree should be included in this universe (Levy, Young, and Zürn 1995). Beyond this lie familiar problems arising from the facts that it is impossible to conduct controlled experiments involving international regimes, and that it is hard to collect comparable data on a large enough sample of cases to allow for application of familiar statistical procedures and other variation-finding techniques of comparative analysis (King, Keohane, and Verba 1994).

In the circumstances, it will come as no surprise that the outcome of efforts to evaluate competing claims of the two families of models is often a kind of stalemate in which those who prefer specific models adhere to their positions as a matter of faith and talk past one another in their efforts to convince others of the validity of their positions. The debate between proponents of the management model and the enforcement model of compliance is a particularly notable case in point (Chayes and Chayes 1995; Downs, Rocke, and Barsoom 1996; Mitchell 1996). But the problem extends to the most basic assumptions of the two families of models. This is an unsatisfactory state of affairs from the point of view not only of scholars but also of practitioners who must make difficult choices both about the design of regimes and about their administration once they become operational. What can we do to alleviate this state of affairs and, in the process, add significantly to our understanding of the roles that regimes can play in solving large-scale environmental problems?

As a first step, I suggest that we launch a search for critical tests of the sort familiar to natural scientists facing similar problems in their research programs. The essential idea is to look for a small number of issues— no more than two or three—with regard to which competing models or theories offer highly different predictions that lend themselves to empirical investigation. Having identified such targets of analysis, it may be possible to structure the incentives of members of the research community in such a way that they will be motivated to compete with one another in devising persuasive answers to the question of which divergent prediction is correct. Handled properly, this can generate both a focused debate and healthy competition in which individual researchers or teams of researchers are rewarded for their efforts to produce and defend their answers to common questions. The entire scientific community stands to gain as a primary beneficiary of these efforts.

Proceeding in this fashion, I propose two critical tests of predictions about international environmental regimes arising from collective-action and social-practice models. My first candidate centers on the relationship between compliance and enforcement. Collective-action models predict that with the exception of regimes dealing with coordination problems where there is no incentive to cheat, violations of rules and failures to fulfill commitments will be common in the absence of one or more well-developed enforcement mechanisms. More specifically, we can expect to find an inverse relationship between levels of compliance in the absence of such enforcement mechanisms and costs to subjects, including both regime members and relevant actors operating under their jurisdiction, of complying with specific rules or fulfilling substantive commitments. In other words, the higher the costs of compliance and the stronger the incentive to cheat, the greater the importance of enforcement. As a corollary, the deeper cooperation goes, the greater the significance of enforcement (Downs, Rocke, and Barsoom 1996).[4] According to social-practice models, there should be no clear-cut relationship between compliance and enforcement. This does not mean that these models predict that compliance with rules and commitments will be high regardless of the presence or absence of effective enforcement mechanisms. Rather, it simply means that there should be no simple and clearcut relationship between compliance and enforcement.

My second candidate to serve as a focus for critical testing centers on the stickiness or persistence of international environmental regimes. Collective-action models predict that regimes are fragile constructs or even that they will emerge as epiphenomena. They often collapse or become, for all practical purposes, dead letters when the bargains that underlie them disintegrate either because of a shift in the distribution of bargaining power or because one (or more) key member state changes its policies regarding support of the regime. Social-practice models predict that regimes can survive and even thrive after such changes. This does not mean that the models predict that institutional arrangements will persist under all conditions. Rather, it means that no systematic relationship should be established between shifts in bargaining strength or interests of major actors and the fate of individual regimes. This condition would be met if a sizable proportion of regimes under consideration survive after major shifts in the political relationships that give rise to them.

What methods are most appropriate to the conduct of such exercises in critical testing? In my judgment, we are unlikely to find any silver bullets in this field of study. No single method can be expected to produce answers to the questions posed above that will prove convincing to unbiased observers. Does this mean we have reached an impasse in our efforts to understand the effectiveness of international environmental regimes? I do not think so. What is required now, in my view, is a research program that features a mix of natural experiments, laboratory experiments, and thought experiments combined with a common commitment to comparing and contrasting the findings flowing from these methods. Although they may not be definitive, findings that reflect a convergence of results arising from the use of several distinct methods certainly merit particular attention.

Despite some obvious limitations, each category of methods holds significant promise. Natural experiments turn on the identification of substantial variance regarding a chosen variable, such as the presence of different types of enforcement mechanisms, among sets of institutional arrangements that are otherwise quite similar (Breitmeier et al. 1996). Ideally, this variance should encompass a number of levels in contrast to a simple dichotomy (e.g., the presence or absence of significant enforcement mechanisms). Laboratory experiments are likely to take the form of simulations, with or without significant roles for human players (Axelrod 1984, 1997). The virtue of such research lies in the opportunity it affords to vary individual factors (e.g., the nature of enforcement mechanisms) in a controlled manner, even though this advantage comes at the price of uncertainty about the extent to which the simulated world maps onto the real world. Thought experiments offer the prospect of a disciplined consideration of the way in which real-world situations would have developed if one or another key feature (e.g., the character of enforcement arrangements) had taken a different form. Although it is not easy to achieve rigor in such efforts, recent literature on the uses of counterfactuals suggests that thought experiments do have a role in larger research strategies (Tetlock and Belkin 1996).

What results are likely to arise from this way of dealing with predictions of collective-action and social-practice models regarding the effectiveness of international environmental regimes? It would be premature to jump to conclusions regarding the answer. I expect the returns to come in relatively slowly, so it is reasonable to expect that a sustained commitment to a research program extending over a period of five to ten years will be required to answer the central question with any confidence. Even so, it is worth posing the following question even at this early stage: How should we react in the likely event that sustained research produces mixed results or generates evidence that supports the predictions of collective-action models and other evidence that favors the predictions of social-practice models?

One response features an effort to synthesize the two sets of models, combining essential elements from each to build new models. The attractions of this approach are obvious. But this sort of synthesis is difficult to achieve, and little progress has been made so far in exploring this option among those interested in international environmental problems. Of interest, however, some students of interactive decision making have begun to think in these terms in work that focuses on other areas or other levels of social organization (Ostrom 1998). Perhaps the most promising initiatives start from comparatively parsimonious assumptions of collective-action models and introduce ideas derived from cultural or sociological analyses. The relatively long tradition of work on ideas of bounded rationality and satisficing belong to this line of inquiry (Simon 1957). So also does more recent work that attempts to "examine the implications of placing reciprocity, reputation, and trust at the core of an empirically tested, behavioral theory of collective action" (Ostrom 1998, 1).

Short of producing an overall synthesis of the two families of models, we can ask questions about the domains of validity of the schools of thought. It makes sense to differentiate three distinct lines of inquiry, each of which is worthy of exploration in considerable depth. One strategy is to launch a search for specific intervening variables that can help to account for mixed results arising from efforts to conduct critical tests. It may turn out, for instance, that high levels of transparency can deter utility maximizers from violating rules, even when substantial gains are to be reaped from noncompliance and when enforcement mechanisms in the ordinary sense of the term are poorly developed. Thus, even utilitarians may be affected by the prospect of disapproval, social opprobrium, or development of a reputation for being untrustworthy. Second, we can ask whether certain identifiable types or classes of situations generate incentives to cheat that are particularly difficult to overcome in the absence of effective enforcement mechanisms. This is the crux of the debate about the prevalence of the tragedy of the commons. In essence, strict utilitarians see little prospect of avoiding the tragedy of the commons in the absence of meaningful sanctions or fundamental changes in incentive structures; those who think more in social-practice terms are unconvinced by the need to resort to coercion in the ordinary sense of the term as a means to avoid the tragedy. A third response centers on the ideas that the role of enforcement mechanisms in deterring violations and the character of sanctions most likely to prove effective differ from one issue to another. Those who draw a distinction between high politics and low politics, for example, often have propositions of this sort in mind.

Naturally, it would be desirable to create a unified theory to explain how or under what conditions international regimes can solve large-scale environmental problems. Work on such a synthesis, treated as a long-term goal, should go forward. In the meantime, however, much useful analysis is to be done in pinpointing conditions under which behavioral mechanisms or pathways highlighted by collective-action and social-practice models are likely to prevail (Young 1999a). Clearer understanding of these conditions would not only deepen our knowledge of the role of institutions in international society, it would also provide a basis for producing results of considerable value to members of the policy community responsible for devising and implementing solutions to large-scale environmental problems.

II
Analytic Frontiers

3

Fit: Matching Ecosystem Properties and Regime Attributes

All accounts of the problem of fit start from the same premise: the effectiveness of environmental and resource regimes or, in other words, the capacity of these arrangements to prevent undesirable environmental changes and to solve environmental problems once they arise is determined in considerable measure by the degree to which they are compatible with the biogeophysical systems with which they interact. We are increasingly aware in this connection that ". . . sustainability requires human systems that are concordant at appropriate scales with the ecosystems to which they are related" (Cleveland et al. 1996, 3). We are becoming accustomed as well to asking "to what extent are the . . . rhythms of natural and social systems compatible, and under what criteria can we make their purposes coincide?" (Holling and Sanderson 1996, 58).

It follows that we should resist temptation to think that one size fits all when it comes to designing regimes to solve a variety of environmental problems. We should be leery of simple generalizations purporting to spell out conditions necessary to achieve sustainable human-environment relations (much less to produce efficient or equitable outcomes) across a wide range of biogeophysical settings. An arrangement that works well in dealing with resilient ecosystems capable of recovering quickly in the aftermath of severe disturbances may produce disastrous results when applied to systems in which even moderate disturbances can trigger cascading changes (National Research Council 1996). An arrangement well suited to minimizing the probability of disruptive changes in ecosystems that are relatively self-contained may generate unsustainable results when applied to more open systems in which exogenous forces are major determinants of robustness or stability.

These examples are helpful in clarifying the meaning of the concept of fit in the context of human-environment relations, but they do not provide a basis for developing a systematic account of the problem of fit. How can we devise a conceptually coherent and analytically tractable approach to understanding the problem of fit? One way is to look for parallel features or attributes of ecosystems and human systems on the assumption that they are comparable in terms of their fundamental structures and processes (Costanza and Folke 1996). This approach, which leads to an analysis of ecological and social stocks, flows, and controls, has much to recommend it as a point of departure for development of a general theory of interactions between these systems (figure 3.1). But it does not lend itself to a sustained examination of the implications of specific ecosystem properties for management arrangements necessary to ensure that human actions are compatible with the achievement of sustainability, much less efficiency or equity, in a variety of biogeophysical settings (Berkes and Folk 1998; Pritchard et al. 1998).

In this chapter, I adopt an approach that is less ambitious yet more likely to yield results whose relevance to a study of the institutional dimensions of environmental change is easy to grasp. The first principal section identifies a number of important ecosystem properties and discusses their implications for the design of environmental regimes to deal with problems in which they figure prominently. This is followed by a section that explores sources of misfits or mismatches between properties of ecosystems associated with various environmental problems and attributes of institutional arrangements or regimes created to deal with them. The final section turns to a consideration of issues relating to efforts to eliminate or alleviate these mismatches.

The principal puzzles arising in this context center on the frequency with which mismatches between ecosystem properties and institutional attributes occur and the reasons why it often proves difficult to close these gaps. I propose that the solutions lie in three clusters of variables: the state of knowledge, institutional constraints, and rent-seeking behavior. An important insight is that revealing and even publicizing the existence of a mismatch in a specific case is not sufficient to eliminate its impact on the course of human-environment relations. The point is not to provide an excuse for fatalism or inaction with regard to the institutional

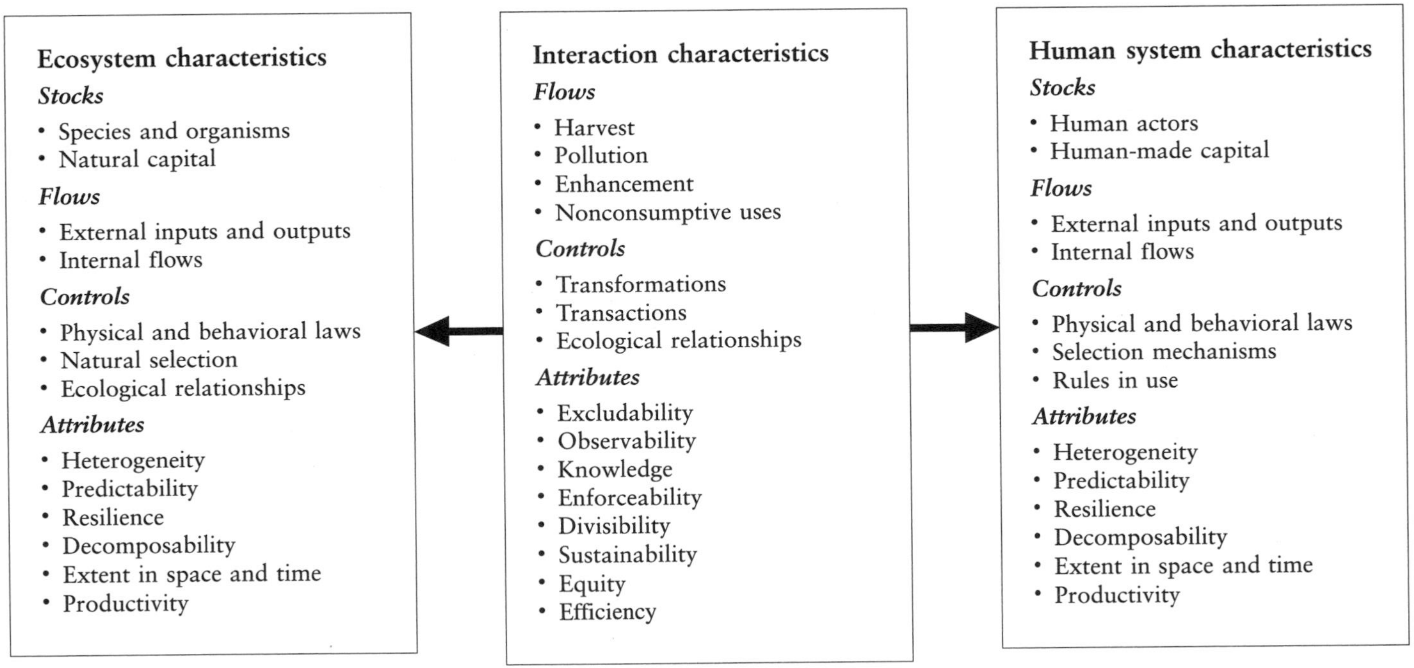

Ecosystem characteristics

Stocks

- Species and organisms
- Natural capital

Flows

- External inputs and outputs
- Internal flows

Controls

- Physical and behavioral laws
- Natural selection
- Ecological relationships

Attributes

- Heterogeneity
- Predictability
- Resilience
- Decomposability
- Extent in space and time
- Productivity

Interaction characteristics

Flows

- Harvest
- Pollution
- Enhancement
- Nonconsumptive uses

Controls

- Transformations
- Transactions
- Ecological relationships

Attributes

- Excludability
- Observability
- Knowledge
- Enforceability
- Divisibility
- Sustainability
- Equity
- Efficiency

Human system characteristics

Stocks

- Human actors
- Human-made capital

Flows

- External inputs and outputs
- Internal flows

Controls

- Physical and behavioral laws
- Selection mechanisms
- Rules in use

Attributes

- Heterogeneity
- Predictability
- Resilience
- Decomposability
- Extent in space and time
- Productivity

Figure 3.1
Ecosystem–human system interactions. *Source:* Young et al. 1999, 51.

dimensions of environmental change. Rather, we are reminded that problems of this sort are ordinarily difficult or impossible to solve through simple exercises in social engineering. They involve political processes, which means that efforts to address them will regularly give rise to institutional bargaining (Young 1994a). In an effort to make these matters more concrete, I turn frequently to cases involving human interactions with large marine and terrestrial ecosystems as examples. But the problem of fit is generic; it occurs whenever and wherever humans interact with biogeophysical systems.

Ecosystem Properties

It is easy enough to point to examples of mismatches between ecosystem properties and regime attributes that have become serious obstacles to sustainability. Particularly striking are cases involving jurisdictional constraints and the use of simple scientific models as management tools. Efforts to conserve highly migratory species (e.g., birds that migrate from one continent to another, fish and marine mammals that migrate from one ocean to another) cannot succeed without cooperation on the part of all states whose jurisdictions they pass through, as well as states exercising jurisdiction over users whose harvesting activities take place in areas located outside the jurisdiction of any state. Similar comments are in order regarding large ecosystems and even larger bioregions that extend across jurisdictional boundaries either at the domestic level (e.g., forest ecosystems in the jurisdiction of two or more states or provinces) or at the international level (e.g., semienclosed seas that are surrounded by two or more coastal states). Although the need to cooperate as a means of achieving sustainability in using the resources of these systems is obvious, conflicts of interest or preoccupation of policy makers with other matters deemed to be of higher priority often complicate or even preclude efforts to design and implement cooperative measures.

Although the mechanism involved is different, the use of simple models as management tools in the administration of regimes can result in equally serious problems. Familiar logistic models used to calculate sustainable yields from stocks of fish offer a clear example. Whatever their merits in analytic terms, incorporation of these models into resource regimes regularly leads to unsustainable harvesting practices in ecosystems featuring assemblages of interactive species and complex dynamics involving physical conditions (e.g., water temperatures) and biological conditions (e.g., reproduction rates) (Larkin 1977). Today, we are aware that marine conservation normally requires a sustained effort to think in terms of large marine ecosystems (Sherman 1992) that sometimes behave chaotically so that relatively small errors in calculations of allowable harvest levels for specific stocks can trigger transformative or cascading changes. The attractions of simple models from the perspective of policy makers and managers are obvious. It is relatively easy to follow standard operating procedures for making collective choices about matters such as allowable catches as long as these models provide the basis for making necessary calculations. But when regimes rely on such models in making decisions pertaining to human activities in highly interactive and chaotic systems, the predictable result is a problem of fit that leads to unintended consequences that threaten or even preclude sustainability (Wilson et al. 1994).

It follows that those endeavoring to create environmental or resource regimes should begin with an assessment of the principal properties of the relevant ecosystem(s) and proceed to design and build institutional arrangements that fit the biogeophysical contours of the problem. Yet this is easier said than done. Even if we set aside the more extreme variants of constructivism, there is something to be said for the views of those who argue that important properties of ecosystems are socially constructed (Jasanoff and Wynne 1998). The idea of an ecosystem is itself an analytic construct useful for many purposes but often difficult to map onto the observable world. A central tenet of modern ecology is the proposition that everything is related to everything else (Commoner 1972). Under the circumstances, specification of boundaries separating distinct ecosystems is ultimately somewhat arbitrary and may emerge as a barrier to addressing important issues. The conventional separation among atmospheric, marine, and terrestrial ecosystems, for instance, can actually impede efforts to understand the dynamics of land-ocean and ocean-atmosphere interactions. The same can be said about the idea that most ecosystems possess stabilizing mechanisms that produce a more or less pronounced tendency to restore the system to some well-defined equilibrium after various types of disturbances (e.g., changes in the abundance of individual species due to human harvesting). Although this idea is attractive in some respects, it masks a striking tendency in some ecosystems to shift from one state to another, and makes it difficult to focus on the chaotic or nonlinear behavior that characterizes many ecosystems, even in the absence of large-scale human interventions.

This discussion should suffice to make us wary of an argument that treats the properties of ecosystems as objective realities or facts of life that are immune from the effects of social construction. Even so, it is hard to deny that ecosystems do have a variety of properties that are more than mere artifacts of human cognition or social discourses and that have considerable relevance for sustainability in human-environment relations. Consider variations in levels of net primary productivity or rates and patterns of regeneration after removals of biomass. It seems clear that regimes governing human harvesting of renewable resources that do not take into account variations among ecosystems in these terms are asking for trouble by ignoring the problem of fit; that is, by failing to devise institutional arrangements that are constructed in such a way that they are compatible with key properties of relevant ecosystems.

The difficulty afflicting efforts to cope with the problem of fit that arises in this connection is quite different from the effects of social construction. No well-specified and widely accepted typology of ecosystem properties exists to anchor a discussion of the fit between ecosystem properties and regime attributes. I am in no position to propose a typology that would win favor among students of biogeophysical systems. Fortunately, however, it is not necessary to solve this problem to proceed with the central thesis of this chapter; a set of illustrative distinctions among familiar categories of ecosystem properties will suffice. In the following paragraphs, I identify a range of ecosystem properties grouped under three broad headings—structures, processes, and linkages (table 3.1)—and comment briefly on their relevance to the problem of fit.

Table 3.1
Ecosystem properties relevant to the problem of fit

Structures
Complexity
Homogeneity
Interdependence

Processes
Productivity
Growth
Stabilization
Change

Linkages
Boundary conditions
Transboundary interactions

Structures

The category of structures encompasses properties pertaining to the elements or parts of ecosystems and the relationships among them. Familiar members of this cluster include complexity, homogeneity, and interdependence. Complexity is a measure of the number of distinct parts or components of an ecosystem and the extent to which the functions of individual components are distinct. Complex systems have large numbers of elements playing functionally distinct roles that are essential to their maintenance. The level of complexity is particularly high when these elements are layered vertically or relate to one another in a hierarchical fashion. A long-standing debate among ecologists concerns the extent to which complexity contributes to the resilience of systems by providing alternative mechanisms for the performance of important functions when individual elements are impaired. With regard to the problem of fit, however, the principal implication of complexity involves increasing problems arising from efforts to compartmentalize occurrences affecting individual elements of ecosystems as the level of complexity rises. Disturbances affecting individual elements of a complex system are likely to ripple through the entire system and may gain force as they move from one element to another, producing what are known as ecological cascades (National Research Council 1996). It follows that regimes created to deal with complex systems must take into account indirect effects of human actions as well as more direct or first-order effects. In situations involving harvesting of renewable resources, for instance, this means considering the effects of harvesting practices on nontargeted species as well as on those intended for consumption.

The same holds true for homogeneity and interdependence. Homogeneity refers to the degree of similarity among individual elements or parts of ecosystems. Homogeneous systems include assemblages of living organisms that resemble one another in biological terms, coupled with physical conditions (e.g., ambient air temperatures, precipitation patterns) that are uniform. Homogeneity may vary greatly from one system to another or even within a single system over time. But in general, the lower the level of homogeneity in ecosystems, the more important it is to devise resource regimes capable of tracking the behavior of various elements of systems individually and making appropriately differentiated responses to developments affecting those elements. In developing management practices to cope with the effects of acid precipitation on lakes, for example, it is important to consider natural variations in the capacity of individual lakes to neutralize acids. In contrast to homogeneity, interdependence refers to the tightness of links or couplings among elements or subsystems of an ecosystem. Elements of an ecosystem may be homogeneous but have a high degree of independence from one another. Conversely, individual subsystems of heterogeneous systems may be tightly coupled. Broadly speaking, the higher the level of interdependence, the more important it is to adopt a holistic perspective or a whole ecosystem approach in managing human actions affecting these systems. In systems where marine mammals depend on the same stocks of fish that are targets of human harvesting, for instance, it is essential to consider these predator-prey relationships in efforts to calculate allowable catches for humans.

Processes

The category of processes refers to ecosystem dynamics that develop, maintain, or transform individual systems through time. Among familiar properties belonging to this cluster are productivity, growth, stabilization, and change. Ecosystems vary greatly with regard to what is sometimes called their metabolism and measured in terms of criteria such as net primary productivity, production of harvestable surpluses, and rates of regeneration after more or less severe depletions of individual elements. These metabolic processes often involve natural cycles whose length, magnitude, and regularity vary from one system to another. The implications of this set of properties for the creation and operation of resource regimes are profound. Management practices that assume the existence of high rates of productivity on the part of targeted species but that operate in settings where productivity is unusually low, can lead to depletions that are so severe that the systems are unable to recover. Much the same is true of assumptions about regenerative processes. As those dealing with marine fisheries have discovered again and again, depletions of targeted species may lead to unintended and undesirable changes in an ecosystem's species composition, even when no net loss of biomass occurs.

Growth, stabilization, and change have related management implications. Whereas growth is a developmental process, stabilization refers to the capacity of an ecosystem to recover from disturbances or to return to some earlier state in the wake of disruptions. It is important to manage human actions in such a way that they do not interfere with normal natural growth cycles. Rules designed to allow escapement of a sufficient number of fish to ensure adequate reproduction are a typical example. With regard to stabilization and change, a major concern arises from the distinction between brittleness, construed as vulnerability of an ecosystem to sudden shocks, and vulnerability of the system to pressures—often referred to as critical loads—that mount slowly over time. Ecosystems that are remarkably resilient under the impact of cumulative pressures may collapse suddenly when hit by sharp shocks and vice versa. A particularly important consequence of this is that resource regimes associated with brittle systems must be sensitive to the prevention of sudden shocks if they are to produce sustainable outcomes. For its part, change may be cyclical, episodic, or chaotic. Ecosystems vary greatly in terms of both the nature of triggers that set change in motion and in the extent to which resultant processes are path dependent or constrained by events occurring within a system during previous times. Relatively crude management systems may produce sustainable results with regard to human activities affecting resilient systems in which change is cyclical and path dependent. But the same arrangements may be disastrous when applied to ecosystems in which modest triggers can touch off nonlinear changes that radically alter the capacity of the systems to sustain human harvesting on a continuing basis.

Linkages

The category of linkages refers to connections between individual ecosystems and their environments (including other ecosystems). What is sometimes called the first law of ecology—the proposition that everything is related to everything else—suggests that there are no completely closed systems, or that exogenous forces will play a role no matter how or where the boundaries of individual systems are drawn (Commoner 1972). Both the strength and significance of linkages between ecosystems and their environments vary greatly. In some cases (e.g., the marine systems of the central Arctic basin), it seems intuitively plausible to assume that biogeophysical systems are relatively self-contained. This initial impression is accurate up to a point; but now let certain long-range transport mechanisms (e.g., airborne or waterborne transport of persistent organic pollutants) or even global processes (e.g., changes in Earth's climate system) come into play. Under these conditions, it quickly becomes apparent that what were relatively self-contained systems are now affected, in some cases dramatically, by these exogenous events. Major changes are occurring in the extent and thickness of sea ice in the Arctic basin. This suggests both that it is dangerous to ignore potential linkages even in seemingly self-contained systems, and that significant linkages may arise over time even where none existed at an initial point.

Implications of linkage properties for the problem of fit are twofold. It is important to exercise care in establishing boundaries of the domains of resource regimes in the first place; a regime that ignores what turn out to be significant elements of an ecosystem cannot produce sustainable results. In addition, it is necessary to review the relationship between ecosystem boundaries and regime boundaries from time to time. A regime whose coverage is perfectly appropriate at the time of its creation may prove inadequate at a later stage in the wake of exogenous developments. What is required in this connection is a practice that involves monitoring and managing boundaries rather than setting them once and for all at the outset.

Sources of Mismatches

Dramatic mismatches between ecosystem properties and regime attributes are all too common. In the American west, to take a striking example, the federal government imposed a system of land tenure intended to guide the settlement of the Great Plains on the arid lands of the Four Corners region where Colorado, Utah, New Mexico, and Arizona come together (Stegner 1954).[1] A study of the dust bowl of the 1930s cited reasons to doubt the appropriateness of this regime, which assumes that a family can sustain itself by working 160 acres of land, even as applied to some parts of the Great Plains (Worster 1979). But the consequence of applying this system to the arid lands of the southwest was an unmitigated disaster. A regime designed for areas receiving an average annual rainfall of twenty inches or more could hardly be expected to produce sustainable livelihoods for residents of areas where average annual rainfall is more like three or four inches. Where the universe of cases is heterogeneous, in other words, one size does not fit all with regard to the performance of institutions.

The history of the regime for whales and whaling offers an equally dramatic example of a mismatch occurring at the international level. Those who established the whaling regime in the 1940s drew on the intellectual capital accumulated in efforts to manage fisheries and assumed that models used to calculate maximum sustainable yields from fish stocks would work reasonably well when applied to the harvesting of whales (Gulland 1974; Small 1971). But whatever the virtues of these models in providing a basis for the management of fisheries (and this is a matter of considerable debate), they proved severely deficient in the context of the whaling regime. Whale stocks are difficult to monitor precisely, and most are characterized by low rates of regeneration.[2] Under the circumstances, managers often fail to detect depletions in a timely manner or take action to deal with them until depletions have reached a level at which stocks are unable to recover or recover so slowly that an indefinite moratorium on harvesting is necessary to guarantee their survival. In many large marine ecosystems, in fact, removal of whales from the biomass allows other species to expand to fill the (temporarily) vacant niche, a development that sharply curtails regeneration of whale stocks. Evidence suggests that this is exactly what happened in the Bering Sea region during the first half of the twentieth century (National Research Council 1996). In effect, high levels of interdependence among individual elements of this ecosystem ensured that major changes in one important element would ripple through the entire system.

What are the sources of such mismatches, and why do they occur so frequently? Such problems are not attributable solely to human stupidity or avarice, although history does record its fair share of these phenomena. An initial examination of the evidence suggests that a number of distinct mechanisms produce institutional misfits. These sources are not mutually exclusive; two or more of them may operate at the same time. It is perfectly possible, therefore, for specific mismatches to be overdetermined. Nonetheless, we can gain insight into their sources or causes by grouping them into three main categories: imperfect knowledge, institutional constraints, and rent-seeking behavior.

Imperfect Knowledge
Efforts to match institutional arrangements governing human actions to the properties of biogeophysical systems cannot succeed in the absence of usable knowledge regarding the ecosystems in questions. Of course, this knowledge need not be rooted in the Western scientific tradition that dominates most thinking about ecosystems today. Informal knowledge of the sort accumulated by indigenous peoples living in close contact with the same ecosystems over long periods of time can play an important role in efforts to adapt institutions, including arrangements rooted in informal social practices in contrast to formal agreements, to the properties of relevant ecosystems (Berkes 1989, 1999). In some situations, even implicit knowledge has a role, in the sense of de facto norms or rules that arise through a process of trial and error rather than a conscious effort to devise management systems to govern human actions involving the use of natural resources (Fienup-Riordan 1990). Much of the burgeoning literature on the development of common property regimes in small-scale systems deals with arrangements that rest on informal or implicit knowledge (McCay and Acheson 1987). But none of this alters the fact that avoiding mismatches requires usable knowledge in one form or another.

We should never underestimate the influence of simple ignorance about the behavior of biotic and abiotic systems as a source of misfits, especially when it is combined with an attitude of dominance that licenses or even encourages human exploitation of natural resources unless and until the consequences become demonstrably destructive. In some cases, ignorance takes the form of lack of awareness or understanding regarding straightforward factual matters. Harvesting often leads to depletions of stocks of renewable resources that are not detected or documented until the problem becomes unusually severe or even unsolvable (Harris 1998). Emissions of pollutants (e.g., ozone-depleting substances) frequently go on for some time before their effects on biogeophysical systems are detected, much less documented in an irrefutable manner (Roan 1989). In other cases, ignorance involves lack of understanding of causal mechanisms at work in large dynamic systems. Even in severely depleted fisheries, for instance, we still have trouble sorting out the relative weight of anthropogenic forces (e.g., overharvesting) and nonanthropogenic drivers (e.g., changes in water temperatures) as causes. And the primitive nature of our grasp of the dynamics of Earth's climate system is clearly one major source of disagreements about how to deal with emissions of greenhouse gases in efforts to create a global regime designed to protect the system.

Ignorance can lead to institutional mismatches in at least two ways. Human users may simply fail to consider key facts or causal mechanisms, concluding that regimes are unnecessary or building regimes that overlook important properties of relevant ecosystems. Arrangements designed to determine allowable harvest levels in specific fisheries without consideration of the dynamics of the larger ecosystems in which individual stocks of targeted species are embedded exemplify this (Sherman 1992). Alternatively, users may resort to familiar but ultimately false analogies in an effort to come to terms with facts or causal mechanisms that are acknowledged to be important but are poorly understood. Assuming that the population dynamics of whale stocks will resemble the population dynamics of fish stocks is an example. Either way, it is easy to see how ignorance can give rise to institutions that are incompatible with important properties of biogeophysical systems affected by human actions.

Important as it is, simple ignorance is not the only link between imperfect knowledge and institutional misfits. To regulate or control the course of human-environment relations, analysts and practitioners regularly construct and apply models to explain or predict the dynamics of important ecosystems. The familiar logistics models developed to calculate sustainable yields from individual stocks of living resources are a particularly prominent example. More broadly, human actors typically develop discourses or ways of framing problems and structuring thinking about them (in contrast to models in the more rigorous sense) that shape efforts to come to terms with problems arising in human-environment relations (Litfin 1994). The tendency to think in terms of equilibrating mechanisms in contrast to nonlinear or chaotic processes in assessing the effects of human actions on ecosystems is a prominent case in point. As long as we assume that a biogeophysical system will exhibit a pronounced tendency to return to some earlier state in the wake of more or less severe disturbances, we have no compelling reason to adopt a precautionary attitude in regulating human actions affecting the system. The point is not that models or, more generally, discourses are bad. In fact, building models can become a powerful tool in the hands of those seeking to understand the dynamics of complex systems. In any case, it is not possible to escape the influence of models and discourses in thinking about the behavior of ecosystems. Rather, the problem is that faulty models or misleading discourses can go far toward producing mismatches between ecosystems and attributes of regimes humans create to govern their interactions with these systems. The fact that specific models or discourses can become entrenched through processes of socialization or development of standard operating procedures on the part of management agencies only intensifies the problem. Once again, logistics models used to support calculations of sustainable yields in various fisheries provide an illustration. We know that these models are inadequate and that the results they produce fail to take critical features of relevant ecosystems into account in many marine settings (Larkin 1977). Yet even today it is difficult to break their grip on the thinking of those responsible for setting allowable harvest levels for specific fisheries on an annual basis.

Imperfect knowledge as a source of institutional misfits takes on added significance when it comes to dealing with human-dominated ecosystems (Vitousek et al. 1997). We are used to drawing a clear distinction between biogeophysical systems and social systems. In the typical case, we assume that it is feasible to understand the dynamics of biogeophysical systems without reference to human actions, and that it makes sense to employ the resultant knowledge as a basis for making decisions about permissible human uses of natural resources or environmental services (e.g., establishment of air quality standards to regulate emissions of various airborne pollutants). Increasingly, however, we are coming to the realization that humans are major (sometimes dominant) players in ecosystem dynamics. Although the extent of human dominance has increased dramatically in the recent past, evidence points to important roles humans have played in ecosystem dynamics over at least the last ten millennia (Turner et al. 1990; Ponting 1992). The significance of this for the problem of institutional misfits is straightforward but important. We have to endogenize the role of human actors to develop models of coupled human-natural systems to be used in efforts to create appropriate regimes. A regime designed to protect biological diversity that does not take into account the role of humans as agents of land-cover change, for instance, cannot be well matched to the problem it is intended to address. Yet endogenizing human actions is easier said than done. As land-cover change makes clear, efforts to incorporate human actions into models of terrestrial ecosystems are still at an early stage. Thus it will come as no surprise that mismatches between ecosystems and institutions attributable to imperfect knowledge are likely to be particularly severe in cases featuring human-dominated systems.

Institutional Constraints

Regimes created to govern human actions affecting ecosystems are typically embedded in larger or overarching social institutions. Of course, this is obvious at the national level where management arrangements pertaining to fish, forests, and pollution are created through legislative processes and entrusted to specific government agencies to implement (Klyza 1996). But similar links exist at other levels of social organization. Individual international regimes dealing with renewable resources (e.g., whales), geographically defined areas (e.g., Antarctica), or pollution (e.g., intentional oil pollution at sea) are embedded in a society of territorially organized states that has no centralized public authority but that features a rising level of interdependence among its members (Bull 1977). This holds true for institutional arrangements operating in small-scale systems (Jodha 1996). In some cases, resource regimes are integral to the larger social system. It would be incongruous, for example, to attempt to describe reindeer-herding societies or whale-hunting societies without reference to social practices governing the herding of reindeer and the hunting of whales (Krupnik 1993). But even in such cases, specific resource regimes are elements in larger sets of social institutions that can be expected to affect the operation of these regimes.

The existence of institutional linkages is a fact of life that is neither good nor bad in itself. Yet it is easy to see that these relationships give rise to institutional constraints that sometimes lead to mismatches between ecosystem properties and institutional attributes. An obvious case, referred to in several illustrations, arises from the existence of jurisdictional boundaries. In today's world, most politically organized units, ranging from local communities through nation states to supranational organizations (e.g., the European Union), are territorially based. The scope of their jurisdiction, including jurisdiction over adjacent marine areas, is the product of a range of political, economic, and cultural forces. It is a rare instance in which ecological considerations played a significant role in determining jurisdictional boundaries. Under the circumstances, it is not surprising that the coverage of regimes frequently fails to match the spatial boundaries of ecosystems. Arrangements dealing with renewable resources that migrate over long distances (e.g., birds) and flow resources that cross the domains of several countries (e.g., international rivers) are obvious examples. But similar problems occur frequently when stocks of nonrenewable resources (e.g., pools of oil) straddle jurisdictional boundaries of two or more political units. It is possible, of course, to address such problems by shifting management authority to a higher level of social organization, as when national governments assume authority to manage resources not confined to individual communities, or by negotiating cooperative arrangements among territorially based units, as when two or more states create international resource regimes. But transaction costs involved in resorting to these strategies are likely to be high. In many cases, unrelated obstacles (e.g., disputes about the locus of jurisdictional boundaries, cultural antagonisms) effectively rule out these kinds of solutions (Lowi 1995). Even though the ecological consequences may be well understood, therefore, we should not be surprised when jurisdictional boundaries cause mismatches between ecosystems and institutions.

Nor are jurisdictional problems the only institutional constraints. Even when attributes of environmental or resource regimes seem well suited to the properties of relevant ecosystems in principle, mismatches may emerge during the course of implementation or the effort to move institutional arrangements from paper to practice (Mitchell 1994). In part, this is a matter of bureaucratic politics (Allison 1971). A number of units may compete for the status of lead agency for purposes of administering a regime, and the winner in bureaucratic terms may have a corporate or collective ideology or a management style that is not conducive to the pursuit of sustainability in managing human uses of relevant ecosystems (Clarke and McCool 1996). Battles between the U.S. Fish and Wildlife Service and the U.S. Geological Survey over the management of lands in Alaska come to mind. Numerous agencies may want a voice in any decisions made about certain ecosystems, but none may be able or willing to take on the role of lead agency. In such cases, rules in use are likely to become haphazard and to bear little resemblance to features of key ecosystems, even when a regime exists on paper that seems reasonably well matched with the relevant ecosystem properties.

Somewhat similar observations are in order regarding the impact of interest group politics, especially in relatively open settings such as the American political system. It is particularly important to note in this connection that most interest groups (e.g., fishers, loggers, oil producers) do not refrain from efforts to influence the character of institutional arrangements once they have been established through legislative action or institutional bargaining. As a result, arrangements that seem reasonably well matched to ecosystem properties on paper may turn out to be poorly matched to these properties in practice. A sizable fraction of the debate about emissions trading in the case of climate change, for instance, turns on disagreements about how carbon markets would work in practice rather than on arguments about results of the operation of ideal carbon markets.

Beyond this lies the institutional problem often described in terms of path dependence; in other words, the pronounced tendency of human systems to follow well-defined courses once they are launched on particular paths. A number of analysts have pointed to a distinction between fast variables and slow variables in thinking about human-environment interactions (Holling and Sanderson 1996). Fast variables, including technological changes as well as many biogeophysical processes, involve features of systems that can experience dramatic, even transformative, changes over short periods of time. Slow variables, such as intentional changes in social institutions, generally unfold slowly. The problem in this connection is easy to identify, even though it is difficult to solve. Many ecosystems undergo rapid changes; cascades leading to shifts in assemblages of species can occur over periods of months to years. Much the same is true of technologies used to harvest renewable resources or implicated in emissions of pollutants. Institutional arrangements, in contrast, often change or evolve at a much slower pace; major adjustments in many although by no means all resource regimes can take years to decades. It follows that path dependence with regard to the behavior of environmental and resource regimes can become a source of more or less serious mismatches between ecosystems and institutions. Sudden changes in fish stocks due to abiotic forces such as shifts in water temperature may not trigger changes in procedures for calculating allowable harvest levels until serious damage is done (Dobbs 2000). Rapid developments in harvesting technologies such as introduction of high-endurance stern trawlers may not be followed by appropriate adjustments in procedures for regulating the actions of harvesters until a severe crisis occurs (Warner 1983). Such mismatches need not be accepted as unalterable. Much interest is being expressed in devising more flexible institutions, capable of monitoring and adjusting quickly to changing ecological conditions. Even so, path dependence on the part of institutions created to deal with highly dynamic ecosystems or socioeconomic systems looms as an important source of institutional misfits in many settings.

Rent-Seeking Behavior

The concept of rent seeking emerged over the last several decades as a way of organizing thinking about tensions or even conflicts between pursuit of individual gains and promotion of social welfare (Buchanan, Tollison, and Tullock 1980; Tullock 1989). In principle, everyone should have an interest in enhancing social welfare, at least in the sense of expanding the size of the common pie in contrast to distributing a fixed pie among individual recipients. As long as each expansion meets the test of Pareto optimality, everyone's lot should improve (or at least remain unchanged) as the social welfare frontier moves in a northeasterly direction.[3] But as most observers now acknowledge, this perspective fails to explain a sizable fraction of situations in which actors try to improve their individual payoffs at the expense of social welfare. This is attributable in part to the contrast between relative gains and absolute gains (Baldwin 1993). An actor whose absolute gains increase as a consequence of group action may nevertheless conclude that his or her situation has deteriorated if others experience greater gains in relative terms. In part, however, the emphasis on social welfare overlooks the central role of classic distributive concerns in contrast to integrative efforts. In effect, actors frequently devote themselves to improving their individual lot without regard to consequences for others, including members of future generations. In some cases, the social consequences of their actions may go unnoticed or be hard to compute in any rigorous way. In other cases, actors may simply be indifferent to the effect of their actions on the welfare of others. In any case, emphasis on the pursuit of individual gains without reference to the implications of such actions for social welfare is increasingly discussed in terms of rent-seeking behavior.

Rent-seeking behavior can lead to mismatches between ecosystems and institutions in at least two major ways. One centers on economic practices often characterized by environmentalists as "rape, ruin, and run." In the absence of rules designed to protect the public interest (e.g., public trust doctrines), private actors often exploit natural resources ruthlessly for their own benefit (Hays 1959). They destroy forests, shifting their operations to new lands when areas subjected to heavy logging are exhausted. They move their places of residence to distant locations so that they are insulated from the effects of pollution in areas affected by activities of industries they work for or own. They transform natural capital into industrial capital with the result that the condition in which ecosystems are left has no direct bearing on their individual wealth or well-being. The mechanisms differ from one case to another, but the underlying problem in such situations is invariably the same. In the absence of rules that compel private actors to pay attention to the welfare of others, there is every reason to expect that rent-seeking behavior will lead to situations in which renewable resources are overexploited for consumption, the use of resources to produce nonexcludable goods is marginalized, and little attention is paid to the effects of (especially long-range) pollution. In every case, these occurrences reflect situations in which institutional arrangements are poorly matched with the properties of the ecosystems they address.

Beyond this, rent-seeking behavior is not limited to situations involving straightforward or conventional economic practices. In most social settings, issues relating to natural resources and the environment are matters of public policy and therefore subject to manipulation on the part of actors seeking to promote their own interests through political processes. The case of land tenure in the Four Corners region of the American southwest is a good illustration. In the period after the Civil War, territories seeking to qualify for statehood were obliged to acquire a minimum population and to demonstrate a capacity to attract additional settlers. Under the circumstances, political leaders hoping to become governors of new states had good reason to attract new settlers, regardless of long-term ecological consequences. One approach was to extend the Homestead Act to these lands, thereby attracting settlers with the promise of land despite the mismatch between ecological conditions envisioned in the act itself and those actually obtaining in the Four Corners region. To cover up this environmentally inappropriate strategy, supporters invented the ingenious but disingenuous and ultimately destructive thesis that rain follows the plow (Stegner 1954). But this could hardly mask, much less alter, the underlying mechanism. A relatively severe mismatch between ecosystem properties and institutional attributes arose as a consequence of political rent seeking on the part of leaders able and willing to pursue their own objectives regardless of longer-term consequences for sustainability of human-environment relations.

Interaction Effects

So far I have described the sources or causes of institutional mismatches on the assumption that the forces at work are independent of one another. This makes sense from an analytic point of view. Because individual sources are relatively complex and poorly understood, the advantages of treating them on their own merits are obvious. Nonetheless, it would be inappropriate to conclude this account without noting that individual causes of mismatches frequently interact with one another. Any number of interactions are possible. It is not feasible to provide a systematic treatment of this topic in this brief account, but a few examples should convey a sense of their significance.

One common interaction involves linkages between rent-seeking behavior and imperfect knowledge. With regard to most major environmental problems, significant uncertainties surround both the behavior of relevant biogeophysical systems and the capacity of humans to cope with the effect of environmental problems in such a way as to avoid serious losses of social welfare. In particularly complex situations, these uncertainties are likely to be profound. In the case of climate change, there is room for serious debate not only about whether we are currently witnessing the onset of global warming attributable to anthropogenic forces but also about the underlying dynamics of Earth's climate system. What is more, scientists are not immune from the influence of a variety of pressures (e.g., promises of enhanced funding for their research) that go well beyond the bounds of scientific reasoning as such (Jasanoff and Wynne 1998). Despite the best efforts of the Intergovernmental Panel on Climate Change to arrive at carefully crafted and rigorously reviewed conclusions about a number of matters pertaining to climate change, the scope of uncertainty regarding this problem remains wide. The result is an obvious opportunity for those benefiting from the current state of affairs to exploit uncertainty to their own advantage. It is no surprise that the energy industry employs and deploys its own experts who assert that the problem of climate change is not something to become exercised about, or that groups of consumers who prefer energy-intensive lifestyles find it easy to discount reports about climate change that make it hard to justify activities such as unnecessary use of sport-utility vehicles. Of course, these skeptics may prove to be right in the long run. But a far more likely outcome will be an increasingly serious mismatch between properties of Earth's climate system and attributes of institutional arrangements governing emissions of greenhouse gases.

A second common type of interaction involving distinct sources of mismatches features links between the influence of dominant discourses or paradigms and bureaucratic politics. Public agencies, especially the more successful ones, generally develop more or less well-defined points of view or ideologies that give them a kind of collective or corporate personality (Kaufman 1960). For agencies dealing with natural resources and the environment, these ideologies are typically rooted in discourses or perspectives on human-environment relations that guide human actions, even in situations characterized by imperfect knowledge (Klyza 1996). In the United States, for example, the Bureau of Reclamation espouses a world view that extols the virtues of producing cheap energy through construction of high dams; the Forest Service subscribes to the doctrine of multiple use, which it interprets as requiring opening public lands for timber harvesting under the terms of long-term contracts; and the Minerals Management Service sees itself as an agency dedicated to opening oil-bearing lands to exploration and potential development on the part of private industry (McPhee 1971). Of course, these agency outlooks are not cast in concrete; they adjust to outside pressures on a time scale measured in decades. Yet it is remarkable how resistant these outlooks often prove, even in the face of mounting evidence regarding the social costs of clinging to the status quo. Explanations for this rigidity include the effects of socialization on agency personnel and the closeness of relations agencies frequently develop with interest groups, especially powerful corporate actors, whose actions they are nominally required to regulate (Stigler 1975). But the main point in this discussion is the tendency of dominant discourses to reinforce the effects of bureaucratic politics, even when mismatches between ecosystems and institutions reach such proportions that they are visible to the general public.

The same can be said about interactions between the effects of cognitive rigidities on the part of individuals and the impact of path dependence as an organizational phenomenon. Although rent-seeking behavior sometimes involves deliberate manipulation, most individuals are socialized to accept certain propositions as articles of faith or convince themselves that the actions they engage in on a day-to-day basis will help advance the common good. There is no reason to question the sincerity of the average oil company executive who dismisses the significance of climate change, the typical bureaucrat who does not question standard operating procedures of his agency, or the ordinary citizen who is loath to accept personal responsibility for climate change. Most individuals have limited tolerance for cognitive dissonance; they adjust conflicting statements of fact, causal inferences, or norms in such a way as to restore harmony in their thought processes. More often than not this means finding some means to minimize the significance of environmental problems that credible observers regard as threats to the maintenance of sustainable human-environment relations. The consequence is a conservative bias in most institutional arrangements governing human activities affecting natural resources and the environment. As a result, mismatches have to become severe and evidence pertaining to them has to be undeniable before action is taken to address incompatibilities between ecosystem properties and institutional attributes. Ironically, these same forces sometimes contribute to the sudden collapse of important institutions. Once the faith of supporters is undermined, the credibility of systems of roles, rules, and relationships can evaporate almost overnight. But until that point is reached, cognitive rigidities generally deflect the forces for change in institutional arrangements and, in the process, enhance the effects of path dependence.

The Persistence of Mismatches

Why do mismatches between ecosystem properties and institutional attributes often persist over long periods of time and prove resistant even to well-informed and well-organized efforts to close the gap? Part of the answer lies in well-known complications associated with collective action (Young 1989). Revised or restructured institutions are likely to be treated as exhibiting characteristics of public goods (i.e., nonexcludability and nonrivalness), a fact that generates incentives for individual members of relevant social groups to behave as free riders when it comes to attempts at institutional reform (Olson 1965). Even when actors do become engaged in efforts to restructure institutional arrangements, moreover, they are likely to bring conflicting preferences regarding options for improvement to the process, so that the pursuit of institutional reform regularly gives rise to protracted and costly bargaining over the relative merits of various institutional attributes (Young 1994a). Important as they are, however, these generic or universal factors cannot account for all of the persistence of institutional mismatches. In specific cases, they may be reinforced by forces associated with one or another of the sources of mismatches discussed above. A few illustrations will lend substance to this proposition.

Ideas are sticky, especially when they give rise to coherent and tractable models and when no obvious alternatives can be adopted once the limitations of old models become apparent. Consider, again, the logistics models that provide the analytic basis for standard calculations of maximum sustainable yields in the realm of fisheries management. The limitations of these models, at least as management tools in most marine settings, have been apparent for some time. Today, we are regularly reminded of the importance of thinking about large marine ecosystems, the dynamics of nonlinear changes or cascades, and the role of humans as major players in these ecosystems in making decisions about allowable harvest levels for individual fisheries (Wilson et al. 1994). In the 1970s, prominent specialists were writing epitaphs for the idea of maximum sustainable yield as a management tool (Larkin 1977). Yet it is hard to eliminate the influence of logistics models in day-to-day decision making on the part of those responsible for administering management regimes in this realm. The models are intuitively appealing and easy to use. No straightforward substitute is available to replace the intellectual capital embedded in them. In any case, it takes time to train a new generation of managers who have internalized current thinking about the properties of larger ecosystems to which fish stocks belong. Without doubt, the stickiness of this way of thinking is one—albeit only one—element of the dramatic collapse of cod stocks of the Northwest Atlantic during the 1980s and 1990s (Harris 1998; Dobbs 2000). The absence or underdeveloped character of analytic alternatives is surely a factor to be reckoned with in coming to grips with the crisis in the world's fisheries that was amply documented over the last two decades (McGoodwin 1990). Nor is anything unusual about fisheries in this respect. Apart from occasional episodes of cognitive revolution, the expansion or redirection of intellectual capital is a slow variable in the context of human-environment relations (White 1967).

The same holds true for the influence of institutional constraints. We are all aware that seemingly entrenched political systems sometimes collapse, leading to dramatic institutional changes over short periods of time. But such events are rare. For the most part, legislative processes and agency practices are highly resistant to change, even in the presence of mounting evidence that existing arrangements are not well suited to dealing with current problems and that serious threats to sustainability, much less achievement of efficiency or equity, have arisen. The struggles of those seeking to reform the U.S. Forest Service with its bias toward consumptive uses of wood, even when this leads to net operating losses, and the National Marine Fisheries Service, with its bias toward short-run needs of commercial fishers, are rich case studies of the significance of path dependence as a determinant of the performance of managers and management systems. Of course, part of the problem arises from the power of special-interest groups that are more concerned with promoting their agendas than with pursuit of the common good defined in terms of sustainability, efficiency, or equity. It is not necessary to rehearse the evidence regarding what is known as the capture of regulatory agencies by groups whose actions they are intended to guide, to understand the relevance of this phenomenon to the problem of fit. But it is important to emphasize the impact of path dependence, quite apart from the influence of special interests. Where ecosystems are highly dynamic and subject to periodic transformative changes, the conservative bias built into most institutions can easily produce mismatches that become more severe with the passage of time.

For its part, rent-seeking behavior is a constant feature of human-environment relations. Rent seekers have no genuine interest in sustainability. If they are able to maximize their private welfare by clear-cutting a forest and investing the proceeds in the stock market, or decimating a fish stock and moving on to exploit some other resource, they can be expected to do so. Rent seeking is a universal phenomenon by no means limited to human uses of natural resources or environmental services. But what makes it a particularly severe obstacle to avoiding or alleviating mismatches in human-environment relations is the fact that the gap between private welfare and social welfare, arising from depletion of common-pool resources, production of externalities, and undersupply of public goods, is unusually large. As many observers have noted, this is exactly why it is necessary to construct effective resource regimes at all levels of social organization. A number of initiatives of this sort have proved quite successful, even at the level of international society where the capacity of regimes to enforce regulatory provisions is severely limited. Nonetheless, we can expect dedicated rent seekers to make an effort to block creation of effective regulatory arrangements and to hamper their implementation once relevant parties have agreed to their provisions. Coping with these obstacles is hard enough in the case of simple problems. But as our experience with climate change over the last ten to fifteen years makes clear, the ability of rent seekers to slow or even block progress is greatly enhanced when it comes to solving large-scale and unusually complex environmental problems.

What is to be done? There are no simple antidotes to these forces leading to the persistence of mismatches between ecosystem properties and institutional attributes. Even so, the situation is far from hopeless. A few examples will indicate the kinds of strategies available. One approach that is practical in a variety of settings centers on systems of implementation review (SIRs) (Victor, Raustiala, and Skolnikoff 1998). The purpose of SIRs is to monitor both the status of key ecosystems and the performance of major environmental or resource regimes. The idea is that continuous, detailed, and credible feedback regarding the course of relevant human-environment relations can reveal mismatches and provide evidence required by those desiring to eliminate or alleviate incompatibilities. Needless to say, evidence regarding the existence of mismatches is not sufficient by itself to ensure success in this realm. Beneficiaries of the status quo sometimes have considerable capacity to discredit or reinterpret evidence generated by SIRs, and reformers may lack necessary political influence to promote their objectives in various policy arenas, even when evidence of mismatches is clear and indisputable. But in many cases, a stream of evidence produced by mechanisms accepted by most members of the relevant community as legitimate or unbiased is likely to emerge, at a minimum, as a necessary condition for the success of reform movements.

Another response is to build substantial flexibility into provisions of environmental and resource regimes. Flexibility can be a double-edged sword in several settings. Rent seekers and powerful interest groups may exploit institutional flexibility to force through changes that promote their own agendas, regardless of the consequences for sustainability, efficiency, or equity. Consider the recent efforts of antifur campaigners

(Lynge 1992). Yet it is apparent that flexibility is often critical to the success of efforts to alleviate or eliminate mismatches. This is particularly true when knowledge regarding dynamics of key ecosystems is limited at the time of regime creation and where the ecosystems are prone to nonlinear and even transformative changes. It is encouraging to note in this connection that such flexibility can be built into regimes even at the international level where sovereignty sensitivity often dictates the adoption of rigid procedures. The rules of the ozone regime allowing for changes in phase-out schedules for certain types of chemicals without ratification by individual member states is a particularly noteworthy example of flexibility at the international level (Gehring 1994). But it is not difficult to find other examples, particularly where regimes rest on informal or soft-law agreements, that offer hope for the role of flexibility as a means of reducing incompatibilities between ecosystem properties and institutional attributes in a variety of fields.

Another approach, which may substitute for SIRs and flexibility, features what is known as the precautionary principle. The basic idea is to respond to problems arising from imperfect information and institutional constraints by erring on the side of safety; that is, by building in margins of safety to ensure that exploited components of ecosystems are not pushed beyond the limits of sustainability. Like other responses to mismatches, this one is vulnerable to exploitation on the part of special interests. An interesting example involves the regime for whales and whaling in which those opposed to any harvesting have sought to make use of the precautionary principle to block efforts to put the Revised Management Procedure into practice or, failing that, to impose such strict requirements under the terms of this procedure that the result is little short of a de facto moratorium (Friedheim 2001).[4] A striking feature of this case is the fact that environmental groups, as opposed to industry, are leading the charge in this manipulative effort. But even so, the idea underlying the precautionary principle as a response to the problem of fit is worthy of serious consideration. Given current limitations on our ability to predict the behavior of complex and often chaotic systems, much can be said for adopting strategies that feature generous safety margins in such forms as restrictive air quality standards or allowable harvest levels. In the end, this strategy offers what is almost always a second-best solution. Still, it may be the best we can do when knowledge is limited and pressures for exploitation are strong.

Conclusion

In one sense, the premise of this chapter rests on a limited and somewhat conservative base. In the long-run, we need a more general theory of human-environment interactions that is capable of making human actions endogenous rather than treating humans as actors located outside the bounds of various ecosystems. But it is likely to be some time before a general theory of this sort arises and is sufficiently well developed to help address the problem of fit. We must also reckon with the element of social construction in concepts and models we have developed to examine ecosystems. To take a prominent example, we are accustomed to drawing relatively sharp distinctions among terrestrial, marine, and atmospheric ecosystems. Yet it is apparent that these distinctions are, in the final analysis, arbitrary and that it is necessary to drop them or at least modify them substantially in coming to terms with many environmental problems. The growing emphasis on developing models of coupled human-natural systems in seeking to understand the global carbon cycle is a dramatic example of this line of thinking.

Still, there is much to be said at this stage for a relatively modest effort to identify a range of important ecosystem properties and focus on the fit or match between them and attributes of institutional arrangements created to govern human interactions with these systems. Glaring examples of misfits, such as regimes that do not cover the full migratory range of wild animals or birds, or the use of equilibrium models in efforts to manage nonlinear or chaotic systems, quickly come into focus in this connection. But only a little probing is necessary to identify a substantial range of mismatches between ecosystem properties and institutional attributes. Given the difficulties confronting efforts to eliminate or alleviate them, the problem of fit will surely occupy our attention for a long time to come. Naturally, the search for a more general theory of human-environment relations must go forward. For those concerned primarily with the institutional dimensions of environmental change, however, the research agenda for the foreseeable future with regard to the problem of fit is clear.

4

Vertical Interplay: The Consequences of Cross-Scale Interactions

The boundaries separating institutional systems, like those between individual ecosystems, are often indistinct and difficult to identify with precision. This is a consequence of not only the effects of functional interdependencies linking institutional arrangements to one another but also the role that social construction plays in determining the scope or domain of individual institutions. As a result, efforts to understand how specific institutions operate at the margins frequently run into trouble. But nothing in these observations alters the facts that human societies at all levels of social organization are more or less densely populated with well-defined and widely recognized institutions organized around a variety of functional concerns and spatial domains. And these arrangements frequently interact with one another producing consequences that are too important to disregard.

Given the complexity of individual institutions, it is easy enough to understand why analysts tend to focus on specific arrangements, asking questions about the formation, performance, and evolution of these systems on the assumption that a consideration of forces exogenous to individual institutions is not essential for these purposes. As the number of institutions operating in a given social space (i.e., institutional density) rises, however, opportunities for interplay between or among institutions increase. In complex societies, institutional interplay is common, and the resultant interactions can be expected to loom large as determinants both of the performance of individual institutions and of their robustness or durability in face of stresses or pressures for change. With regard to institutions that address environmental matters or what are commonly referred to as resource or environmental regimes (Young 1982), interplay is a force to be reckoned with in evaluating whether regimes produce results that are sustainable, much less outcomes that meet standards of efficiency and equity.

To prepare the ground for the analysis to follow, several additional distinctions are worth spelling out. Whether the links are vertical or horizontal, individual cases of institutional interplay range along a continuum from essentially symmetrical to largely unidirectional interactions. Symmetry in this context increases to the extent that the effects of institutions on one another are reciprocal. Reciprocity is a common feature of interactions between economic and legal-political arrangements. Economic activities generate resources necessary to operate political systems, but regulatory arrangements impose significant restrictions on the conduct of those engaged in these transactions. Asymmetrical linkages arise when the operation of one institution affects others significantly without triggering equivalent responses. Asymmetry is common where global or regional arrangements produce initiatives, such as the European Union's ban on importation of certain seal products, that have dramatic consequences for local institutions (e.g., mixed economies of remote communities in Greenland and the eastern Canadian Arctic) that are unable to respond in an effective way (Wenzel 1991). Of course, symmetry in this connection is a variable. It can range from high to low across the universe of cases in which institutional interplay is a significant factor, and it can change over time in specific interactions.

With respect to vertical interplay in particular, it is useful to distinguish between interactions involving adjacent institutions and those that involve more remote arrangements. The most common forms of vertical interplay undoubtedly feature links between arrangements that deal with related issues and that are located at adjoining levels of social organization. Prominent cases in point are interactions between federal and state-provincial regimes dealing with management of land and natural resources, and between state-provincial and local arrangements addressing matters of environmental quality or public health. But globalization has prompted growing awareness of the significance of linkages between more remote levels of social organization. It is increasingly apparent, for instance, that developments centered on (re)formation of global rules pertaining to harvesting of marine mammals or tropical timber can have profound consequences for the sustainability of social practices operating at the level of small-scale, local communities. In the aggregate, moreover, outcomes flowing from the operation of many small-scale systems can generate cumulative consequences that are globally significant with regard to matters such as loss of biological diversity. As a result, interest in the nature and consequences of remote linkages is on the rise, although adjacent linkages are likely to remain the primary focus of attention among those seeking to understand the dynamics of vertical interplay.

Vertical interplay encompasses a sizable family of interactive situations. No generally accepted taxonomy exists at this stage, and this chapter does not attempt to devise one. Rather, the goal is to illuminate the general character of vertical interplay by examining in some depth interactions across several levels of social organization relating to land and sea use. The emphasis is primarily, although not exclusively, on vertical interplay arising from functional interdependencies, in contrast to the deliberate or intentional links associated with the politics of design and management. The latter, which often become prominent in connection with efforts to solve problems arising from the effects of functional interdependencies, are a separate theme and form the central topic addressed in the next chapter's analysis of horizontal interplay. Vertical interplay involving political design and management and horizontal interplay arising from functional interdependencies are obviously important in their own right. But in the interests of examining several specific types of interplay in depth and exploring the phenomenon in generic terms, I devote less attention to these forms in this chapter and the next.[1]

To be more explicit, this chapter analyzes environmental consequences of vertical interplay in two relatively familiar settings. The first section examines interactions between (sub)national arrangements pertaining to large marine and terrestrial ecosystems and small-scale or local arrangements featuring (often) informal practices involving systems of land tenure and sea tenure. This is followed by a section that focuses on interactions between international regimes dealing with large marine and terrestrial ecosystems and parallel arrangements operating at the level of individual member states.[2] The principal conclusion is that achieving sustainability in human-environment relations requires a commitment to creating arrangements that can manage functional interdependencies on a continuing basis rather than an exercise aimed at selecting the proper level of social organization at which to respond to particular problems. Whereas shifting to a higher level of social organization may have merit as a means of coming to terms with some environmental problems involving large marine and terrestrial ecosystems, it makes sense to be wary of the pitfalls associated with formation of regimes based on this strategy.

Interplay between (Sub)National and Local Resource Regimes

Patterns of land use and the sustainability of human-environment relations associated with them are determined, in considerable measure, by the interplay of (sub)national—predominantly modern and formal—structures of public property and local—largely traditional—systems of land tenure. For their part, patterns of sea use and the condition of the relevant marine ecosystems are affected greatly by the interplay of (sub)national regulatory systems legitimized by the creation of exclusive economic zones (EEZs) during the 1970s and 1980s and subsistence or artisanal practices guiding the actions of local users of marine resources. As a basis for analysis, this section explores the following hypotheses about these interactions. National arrangements afford greater opportunities to take into account the dynamics of large marine and terrestrial ecosystems (Sherman 1992). But regimes organized at the national level also facilitate and sometimes promote commodification; that is, large-scale, consumptive, market-driven, and often unsustainable uses of targeted resources (e.g., forests, fish). They provide arenas in which the interests of powerful, nonresident players generally dominate the interests of small-scale, local users. Local systems, in contrast, favor small-scale uses of living resources that evolve over time from the experiences of resident harvesters, are less tied to market systems, and accord higher priority to sustaining local ecosystems over the long term. Because traditional local and modern national systems commonly coexist, although they seldom enjoy equal standing in relevant political and legal arenas, actual patterns of land use and sea use are affected substantially by cross-scale interactions between these systems. The account of such interactions here draws on examples relating to southeast Asian forests, grazing lands in the Russian north, and fish stocks in the eastern Bering Sea to ground the analysis empirically. But similar forms of interplay involving a wide range of marine and terrestrial resources occur in many other settings.

Systems of Land Tenure

The authority of national governments to exercise jurisdiction over all lands and natural resources located within the boundaries of states in which they operate is widely acknowledged.[3] This is what accords governments the right to promulgate regulations applying to activities of both owners of private property and users of common property. But beyond this, governments often assert far-reaching claims to the ownership of land and associated natural resources in the form of public property by virtue of conquest (e.g., Russian ownership of Siberia), exercise of royal prerogative (e.g., establishment of crown lands in Sweden), purchase (e.g., acquisition of Alaska by the United States), inheritance (e.g., Canada's inheritance of crown lands under the British North America Act of 1867), succession (e.g., Indonesia's claims to lands once owned by The Netherlands in the East Indies as an element in decolonization and acquisition of independence), or some combination of these. In most countries, claims to public property are remarkably extensive. Private property is nonexistent in Greenland. Despite the publicity surrounding privatization, the government of the Russian Federation still claims most of the land base of Russia as public property. The government of Canada treats the lion's share of the country's land base as public property.[4] Even in the United States, which is widely regarded as a bastion of private property and free enterprise, the federal government alone claims almost one-third of the nation's land as public property (Brubaker 1984).

Yet this is not the whole story with regard to systems of land tenure. Although effective control flowed steadily toward national governments during most of the modern era, many small (often indigenous) groups residing in states and engaging in distinctive social practices have not relinquished their claims to ownership of large tracts of land and natural resources (Berkes 1989; Bromley 1992). Often, these claims overlap or conflict with assertions on the part of (sub)national governments to the effect that the areas in question are part of the public domain. Indigenous land claims in British Columbia, for instance, cover virtually all of the province. In some cases, national governments have recognized these claims and taken steps to reach settlements with indigenous claimants. Particularly noteworthy are comprehensive settlements the government of Canada has negotiated with northern indigenous peoples over the last several decades, and the cooperative arrangements under which the government of Denmark and the Greenland Home Rule handle matters pertaining to use of land and natural resources in Greenland. In other cases, efforts of local communities to assert ownership—or even use—rights have met with strong resistance on the part of national governments. Opposition to the efforts of Sweden's Sami to gain recognition of their rights relating to grazing lands is a striking example (Svensson 1997). In still other cases, national governments have made little effort so far to take claims of local communities to rights involving common property seriously. Throughout much of the Russian Federation, where the legacy of collectivization introduced during the period of Soviet rule remains strong, serious land claims on the part of local peoples are just beginning to surface (Fondahl 1998).

How can these clashes between claims of national governments to public property and local claims to common property be resolved? In some cases, such as the settlement of Native land claims in Alaska, the eventual outcome took the form of a formal transfer of title to some lands to Native peoples (or organizations acting on their behalf), usually in return for acceptance on the part of these peoples of the extinguishment of residual claims to other areas.[5] As experiences in places such as Canada, Greenland, and Fenno-Scandia make clear, however, the concept of property encompasses a bundle of rights, and the contents of the bundle can be allocated in a variety of ways.[6] This has given rise to lively debates about the nature and extent of usufructuary rights where user groups have not been granted full title to land and natural resources. Among the most significant issues of this debate is the right of national governments to authorize consumptive uses of forests, hydrocarbons, and nonfuel minerals in areas that are important to the conduct of traditional subsistence activities featuring the harvesting of living resources on the part of local peoples.

What difference does interplay between national systems of public property and local systems of common property make with regard to overall patterns of land use and to sustainability of human-environment relations? The answer emerges from consideration of differences in the incentives of national policy makers and local decision makers. For the most part, governments can be expected to look on public property as a means to promote the national interest through activities inspired by the search for export-led economic growth and the effort to attract foreign direct investment. More often than not this means treating forests and nonrenewable resources as commodities to be harvested or extracted to meet the demands of world markets for raw materials. Two other factors reinforce this approach to the use of public property, especially in the developing world and countries in transition. National governments tend to cater to the interests of politically powerful individuals who have no roots in local areas and who consider concessions covering natural resources located on public property primarily as a means to amassing personal wealth. A particularly virulent form of this phenomenon is crony capitalism and the emergence of black markets that many observers of southeast Asia have described in detail (Dauvergne 1997a). International bodies (e.g., multilateral development banks) whose mandates emphasize acceleration of economic growth in developing countries have often acted to reinforce biases against the preferences of local peoples with regard to patterns of land use (Lipschutz and Conca 1993). The World Bank's support of large-scale irrigation systems, road construction, and extraction of nonrenewable resources throughout the developing world is an example.

It would be a mistake to assume that practices of local peoples do not cause major changes in ecosystems. Ample evidence shows that swidden agriculture, deliberate burning of forest understory, and harvest of wildlife can produce major environmental changes (Krech 1999). Yet as long as their traditional socioeconomic practices remain intact, local peoples do not have strong incentives to harvest timber for export, to extract hydrocarbons or nonfuel minerals to sell on world markets, or to build massive dams to support large irrigation systems and industrial agriculture.[7] Where systems of common property controlled by local users prevail, therefore, we can anticipate that patterns of land use will differ markedly from those likely to arise where systems of public property prevail. In essence, we should expect to find a pronounced tendency toward large-scale exports of products such as timber, palm oil, hydrocarbons, and nonfuel minerals in systems where public property arrangements govern the use of land and natural resources. Local users operating under common property systems are more likely to use land to support subsistence lifestyles and to avoid extraction of raw materials. Naturally, things will be more complex in those increasingly common situations in which the balance between claims to public property and claims to common property is contested or in which efforts to resolve such contests result in complicated and sometimes confusing allocations of the full bundle of property rights among several different groups of claimants.

To see how these interactions play out in practice, consider developments affecting the forests of Southeast Asia and the grazing lands of northern Russia. As a number of observers have pointed out, tropical forests of Indonesia, Malaysia, and the Philippines have been harvested in an unsustainable fashion over the last several decades (Peluso 1992; Dauvergne 1997a). For example, ". . . loggers have degraded much of Southeast Asia's old-growth forests, triggering widespread deforestation" and these activities ". . . irreparably decrease the economic, biological, and environmental value of old-growth forests" (Dauvergne 1997a, 2). Why did this happen? Many commentators have emphasized demand-side considerations, pointing to the role of Japan as a consumer of tropical timber and stating that Japanese companies have few incentives to promote sustainable uses of Southeast Asian forests. At least as important, however, are supply-side considerations and, more specifically, rules governing decisions about alternative uses of the forests. A critical link lies in the creation of systems of public property controlled by newly independent national governments as part of decolonization and establishment of independent states in Indonesia, Malaysia, and the Philippines after World War II. In effect, emergence of public property in these countries made possible the forest degradation that has spread throughout this region. Nothing in such arrangements compels national governments to negotiate forest concessions in the quest for export-led growth and to condone crony capitalism. But the shifting balance between national systems of public property and local systems of common property has played a key role in allowing these developments to occur, since local users pursuing traditional lifestyles have strong incentives to avoid strategies leading to forest degradation and, in the process, undermining the resource base required to sustain these lifestyles. Among other things, this explains the views of many activists who see links between campaigns to reform land use practices that cause forest degradation and the struggle to strengthen the rights of indigenous peoples residing in rural areas of countries such as Indonesia and Malaysia.[8]

A somewhat different illustration involves land use in northwestern Siberia where world-class reserves of oil and especially natural gas were discovered in areas that indigenous peoples, such as the Nenets living on the Yamal Peninsula and the coastal plain of the Pechora River basin, traditionally used as communal migration routes and pastures for reindeer (Osherenko 1995). During the Soviet era, there was little doubt about the choice between hydrocarbon development and protection of traditional lifestyles in this region. The national government claimed ownership of the land and natural resources as public or state property; oil and gas development was granted priority not only as a means to promote economic development but also as a source of hard currency earnings; and concerns of the indigenous peoples were generally ignored or treated as secondary. At the time of its demise, the Soviet Union was the world's largest producer and exporter of natural gas. Yet the 1990s witnessed new developments in land use in this region (Osherenko 1995). This is partly a consequence of the collapse of the Soviet Union and the economic decline occurring throughout the Russian Federation. In part, however, it reflects a growing effort on the part of indigenous peoples to reclaim reindeer from the collective and state farms of the Soviet era, and to reassert common property rights to migration routes and grazing lands necessary to sustain local economies. From the perspective of these peoples, this pattern of land use is superior to nonrenewable resource development, regardless of world market prices for oil and natural gas (Golovnev and Osherenko 1999). It is far too soon to make predictions about what the future will bring in this region. Development of gas fields on the Yamal Peninsula, for example, is in a state of suspended animation. A revival of the overall Russian economy could well generate pressure to resume construction of gas fields and transportation corridors in this sensitive area. But it is clear that the shifting balance between public property and common property will be important in determining future patterns of land use in northwestern Siberia.

Systems of Sea Tenure

The story of sea tenure differs, often quite dramatically, from the preceding account of land tenure. Whereas we have no difficulty organizing our thinking around patterns of land use and systems of land tenure, comparable phrases relating to marine resources—"sea use" and "sea tenure"—have an odd ring. Why is this the case? Broadly speaking, it is fair to say that this divergence stems from the fact that we have little experience with private property and only a limited history of public property in the ordinary or normal sense of the term when it comes to management of human uses of marine resources.

Part of the gap between arrangements dealing with land use and their counterparts governing sea use is attributable to the fact that it is often difficult and sometimes nearly impossible to apply effective exclusion mechanisms to marine resources. Marine resources run together in a fluid manner and, in the case of living resources such as fish, often feature organisms that move freely from place to place in ways that would frustrate efforts to establish possessory rights of individual owners. Seeking to create private property rights in many fish stocks would be like trying to turn flocks of migratory birds into private property in systems of land tenure. Even so, it would be a mistake to exaggerate this feature of property rights in marine resources. Where relevant resources are sedentary (e.g., clam and oyster beds), people have a good deal of experience with property rights, especially use rights that allow holders to exclude others from harvesting these resources in designated locations. Even more highly developed are rights accorded to those who engage in various forms of aquaculture that depend on the existence of secure rights to fish pens and other well-defined marine structures. As these last observations suggest, it is important to consider arrangements under which individual elements in the bundle of rights associated with property come into play, even when little prospect exists for establishing systems based on the full bundles of rights we ordinarily have in mind in thinking about private and public property. In many instances, use rights to particular fish stocks have been established in forms such as preferences granted to harvesters using particular locations and specific types of gear, or rights to harvest a specified proportion of the total allowable catch established for a specific fishery in any given year. The recent growth of systems of individual transferable quotas (ITQs) in a variety of fisheries is particularly noteworthy in this connection (Iudicello, Weber, and Wieland; 1999; National Research Council 1999a).

In part, the scarcity of systems of private and public property associated with marine resources arises from restrictions on the authority or capacity of states to exercise jurisdiction over marine systems. From the beginning of the modern system in the seventeenth century, states have been treated as territorial units possessing virtually unlimited jurisdiction over terrestrial ecosystems located within their borders but very limited jurisdiction over adjacent marine systems (Anand 1983). Early on, states began to assert some jurisdiction over waters located adjacent to their coasts in the form of a narrow belt known as the territorial sea. For the most part, however, granting of jurisdiction over the territorial sea was justified as a requirement for security. Under this arrangement, coastal states agreed to allow outsiders to engage in a variety of activities—innocent passage of ships, laying of submarine cables, overflight by aircraft—taking place within or affecting their territorial seas. Beyond this belt, it was impermissible or inappropriate to lay claim to marine systems as public property in the sense of areas actually owned by the state in the same way that the state owns the public domain.

Given this background, it makes sense to look upon the twentieth century as marked by expansion of the jurisdiction of coastal states over marine systems in both spatial and functional terms (Juda 1996). The three-mile territorial sea grew to twelve miles, and the establishment of exclusive economic zones (EEZs) granted coastal states jurisdiction over approximately 11 percent of the world ocean and most marine living resources. Justified largely on the basis of conservation or achievement of sustainable use, the expanded jurisdiction of coastal states over marine systems now extends to management of a range of activities dealing with harvesting of both renewable and nonrenewable resources and with protection of marine systems from various forms of pollution. Even so, it is important to note that the jurisdiction of coastal states over adjacent marine systems still falls short of the bundle of rights that states exercise over terrestrial systems within their borders. Coastal states do not have the authority to transfer title to marine systems per se to private owners in the way that states have traditionally been able to dispose of sizable portions of the public domain. In many states, it is considered inappropriate even for governments to collect economic returns from the use of marine resources treated as factors of production, a practice that is considered routine in situations involving the use of natural resources (e.g., timber, hydrocarbons) located on the public domain. These restrictions have not deterred states from developing regulatory regimes operated by government agencies (or their subunits) and intended to ensure that users of marine resources pay attention to matters of sustainability and environmental quality. Nor have they precluded introduction of ITQs in which states create rights to harvest specific fish stocks. Nonetheless, they have produced a situation in which it seems awkward to think in terms of systems of sea tenure.

At the same time, substantial parallels can be seen between systems of land use and systems of sea use when it comes to the operation of small-scale, traditional arrangements, quite apart from the aggregation of management authority in the hands of the state. In virtually every case, these local arrangements can be thought of as featuring some form of common property (Pinkerton 1989). It is no surprise that numerous variations occur, depending on the character of the biogeophysical systems, the nature of harvesting procedures, and the content of cultural norms operative among members of the group of appropriators. Nonetheless, almost all these systems have a number of features in common. Although they do not assign full bundles of rights to individual users, they often grant individuals priority in the use of particular fishing sites or of specific gear. They typically exclude outsiders, or nonmembers of the relevant group or community, from using the resources. They normally feature informal arrangements that evolve on the basis of trial and error and that undergo de facto adjustments over time as a way of adapting to changing conditions in the biogeophysical systems or changing circumstances of societies within which they operate. Yet the rules that make up these institutional arrangements are well understood by members of the user communities, and they are buttressed in most cases by compliance mechanisms that are effective in bringing the behavior of individual appropriators into conformance with them.[9]

How have these traditional arrangements governing the use of marine resources performed in practice? As in the case of systems of land tenure, it would be a mistake to idealize indigenous or artisanal systems. To be sure, anthropologists have documented a sizable number of cases in which these local systems have proved sustainable over relatively long periods of time. A particularly intriguing feature of these studies concerns compliance mechanisms (e.g., arrangements featuring taboos) that are effective from the point of view of guiding the behavior of users toward sustainable practices, even when they are not based on scientific understanding of the dynamics of the ecosystems (Fienup-Riordan 1990). Nonetheless, there is no basis for assuming that all traditional systems of sea tenure produce results that are sustainable. Although this is a sensitive and in some circles contested matter, undoubtedly the actual record associated with traditional systems of sea use features a fair number of failures as well as successes, especially in cases involving volatile biogeophysical systems that undergo nonlinear changes from time to time.

By the same token, the record compiled by regulatory regimes created by (sub)national governments to guide uses of marine resources is generally unimpressive. Justified in large part by the need to manage large marine ecosystems on an integrated basis and to bring to bear insights of science to ensure sustainability, these regimes have proved insufficient to prevent a growing crisis in many of the world's fisheries brought on by excess harvesting and inability, both scientifically and politically, to establish and enforce appropriate quotas or other restrictions (McGoodwin 1990). In fact, national governments have regularly provided subsidies to harvesters in a manner that has led to larger and more powerful harvesting capabilities together with heavy debt loads. As this suggests, regulatory regimes established by national governments have a marked tendency to favor the interests of some types of users over others. Thus, large, well-financed, and politically active harvesters have generally profited from the introduction of national systems of sea use, in contrast to small-scale subsistence or artisanal harvesters with little experience beyond the local level and few resources to influence national (or even subnational) policies.

Overall, it seems fair to say that the result has been commodification of marine resources favoring large commercial operators over small operators. This has caused an erosion of the role of traditional common property approaches to sea tenure and led to outcomes that are hard to defend in terms of conservation or even efficiency. National regulators have begun to experiment with a range of policy instruments (e.g., permits to fish, individual transferable quotas) intended to eliminate or suppress some of the worst features of commodification (Iudicello, Weber, and Wieland 1999). The track record of these efforts is not yet extensive enough to justify firm conclusions. Taken together, it is probably accurate to conclude that these institutional innovations show considerable promise at least as responses to the specific problem of overharvesting (National Research Council 1999a; Hanna et al. 2000). Yet we have no basis at this stage for granting high marks to state-based systems of sea tenure with regard to outcomes that are sustainable over time, much less results that can be defended on grounds of efficiency or equity.

To see how the interplay between modern national and traditional local systems of sea tenure plays out in practice, consider the situation that has developed in the eastern Bering Sea region (National Research Council 1996). During the 1970s, the state of Alaska instituted a limited-entry regime for inshore fisheries of this area (fisheries taking place within a three-mile belt over which the state has jurisdiction) largely in response to declining harvests of salmon (Young 1983). Shortly thereafter the federal government followed suit by creating a fishery conservation zone (FCZ) together with a set of regulatory arrangements dealing with harvesting of all species of fish in an area extending from the outer boundaries of state jurisdiction to a point 200 nautical miles from the coastline (Young 1982). Although it would be unfair to state that these initiatives have had no positive consequences, they have given rise to a number of unintended side effects due largely to problems of interplay with other institutional arrangements. The limited-entry system disrupted traditional arrangements featuring a fluid mix of subsistence and commercial harvesting, placed severe restrictions on the actions of young people unable to afford the price of a permit to enter the fisheries, and led to loss of permits among rural fishers whose financial insecurity caused them to succumb from time to time to the temptation to sell the permits to meet short-term needs for cash. For its part, creation of the FCZ precipitated a dramatic rise in participation of American fishers in the eastern Bering Sea and consequent phasing out of foreign fishers. Because the regime established to regulate fishing in this area has the status of a national arrangement, Alaska has been barred from instituting measures to protect local fishers from competition on the part of large, heavily capitalized fishers based in Washington and Oregon. Exclusion of foreign fishers from the FCZ led them to shift their focus to an area of the central Bering Sea located just outside the FCZ and known as the doughnut hole.[10] By the early 1990s, the pollock stocks in this area had collapsed.

During the 1990s, both the United States federal government and the state of Alaska took some steps to address these unfortunate side effects. These include creation of community development quotas (CDQs) intended to bolster the economies of small coastal communities (National Research Council 1999b) and negotiation of a six-nation convention to address the problem of overharvesting of pollock in the central Bering Sea (Balton 2001). Although these are clearly steps in the right direction, it is premature to conclude that they will solve the problems arising from institutional interplay in the Bering Sea region. The CDQs do not provide a substitute in sociocultural terms for a strong cadre of individual fishers, and pollock stocks of the doughnut hole have not recovered sufficiently to activate management procedures established under the 1994 convention. Accordingly, there is a real danger that the innovations of the 1990s will be assessed in the future as responses that were too little and too late. In any event, it is clear that the growth of coastal state jurisdiction over marine resources and emergence of (sub)national systems governing sea use triggered new forms of institutional interplay. The consequences have proved costly not only for many individuals but also for the welfare of small coastal communities in an area such as Alaska.

Interplay between International and National Resource Regimes

Turn now to institutional interplay occurring at higher levels of social organization and, more specifically, to the hypothesis that the effectiveness of international resource regimes—measured in terms of efficiency and equity as well as sustainability—is determined in considerable measure by interactions between rules and decision-making procedures articulated at the international level, and the political, economic, and social systems prevailing in individual member states. International regimes normally set forth generic rules applicable to all members, but leave implementation for the most part to public agencies and other actors in individual member states.[11] But member states often have preexisting domestic arrangements pertaining to specific issues at hand. In any case, they have broader political and legal systems adapted to their own circumstances. As a result, outcomes arising from the operation of international resource regimes will be sensitive to interactions with national arrangements; performance is likely to vary substantially from one member state to another. After an account of the logic of this hypothesis, this section explores cases relating to regimes applicable to tropical timber in southeast Asia and protected natural areas in the circumpolar North, and to regimes dealing with fisheries of the Barents and Bering Seas. As in interplay between national and local institutions, similar dynamics occur in many other settings.

Competence, Compatibility, and Capacity

It is tempting to assume that states that sign and ratify conventions or treaties establishing international regimes will carry out the obligations assumed under these agreements as a matter of course. But as numerous studies of national implementation of international obligations show, there is no basis for such an assumption. Implementation typically varies greatly from one regime to another as well as among individual members of the same regime. As a result, examination of factors influencing implementation at the national level is an important area of regime analysis (Underdal 1998; Weiss and Jacobson 1998; Victor, Raustiala, and Skolnikoff 1998; Underdal and Hanf 2000). What factors determine whether members succeed in implementing the rules of international agreements within their own jurisdictions and whether they accept the results of decision-making procedures operating under the auspices of international regimes? In some cases, this is essentially a matter of political will. Governments sign agreements they have no intention of implementing; executive branch officials who sign international agreements in good faith may be unable to persuade legislators to pass implementing legislation or to allocate necessary resources to operate these arrangements, and changes in the composition of governments can bring to power officials who did not participate in the creation of a regime and have little interest in fulfilling its obligations. At the same time, three factors that are more general and that bear directly on institutional interplay have emerged as important in this context. For shorthand purposes, I call them competence, compatibility, and capacity.

Competence is a matter of political and legal authority necessary to implement commitments made at the international level. In this sense it is largely a function of constitutional arrangements prevailing in individual states. In the United States, for instance, most international conventions do not become legally binding until they are ratified by a two-thirds majority in the Senate. Even then, the Constitution offers no guarantee that commitments embedded in legally binding conventions will always take precedence over domestic laws (Higgins 1994, chapter 12). As a result, American negotiators in international forums frequently oppose otherwise attractive institutional arrangements on the grounds that they probably cannot survive domestic legal and political processes. Small wonder, then, that those representing other countries frequently view the United States as a difficult partner when it comes to creating and implementing international regimes. In other cases, the problem arises from allocation of authority between national and subnational units of government in contrast to separation of powers among components of national governments. In the Canadian confederation, where authority over many issues resides with the provinces rather than the federal government, the government in Ottawa lacks competence to enter into legally binding commitments at the international level regarding many issues without explicit consent of individual provinces.

Compatibility is a matter of the fit between institutional arrangements set up under provisions of international agreements and social practices prevailing in individual member states. Whereas competence is a matter of authority, compatibility concerns standard practices or procedures for handling governance issues that grow up in political systems over time. Given the character of international society, it is generally agreed that member states should be free to implement international commitments within their own jurisdictions in whatever way they choose. But this does not eliminate the problem of institutional fit. Consider a case in which an international regime calls for establishment of a system of tradable permits (e.g., for exclusive use of bands in the electromagnetic spectrum, for extracting minerals from specific sites on the deep seabed, for emitting specific quantities of greenhouse gases), whereas social practices in some member states are based on command-and-control regulations offering little or no scope for incentive mechanisms associated with tradable permits. To make this concern more concrete, think of issues coming into focus with regard to allocation of slots in the geostationary orbit or bands in the electromagnetic spectrum. For those committed to the proposition that tradable permits are essential to ensure efficiency and therefore to secure widespread acceptance of arrangements governing use of these resources, the virtues of allowing and even promoting markets in licenses to use slots and bands seem beyond doubt. Yet such mechanisms are alien to the political cultures of many countries, and government agencies in these countries lack experience with the mechanisms that would allow them to assimilate such a governance system into familiar and well-understood ways of doing business (Chertow and Esty 1997).

Capacity is a measure of the availability of social and institutional capital as well as material resources necessary to make good on commitments entered into at the international level (Chayes and Chayes 1995; Keohane and Levy 1996). We are used to paying attention to capacity when economic and political systems of developing countries lack resources to shift to alternative technologies (e.g., substitutes for ozone-depleting substances) or to enforce international rules within their jurisdictions (e.g., rules pertaining to trade in endangered species) (Gibson 1999). But issues of capacity also arise in connection with the actions of advanced industrial countries. In the United States international commitments are frequently treated with benign neglect when no individual agency is willing to take responsibility for their implementation (i.e., to become the lead agency) or responsible agencies are unable or unwilling to obtain material resources required to play this role. Consider the contrast between American participation in the regime for Antarctica where there is no doubt about the role that the National Science Foundation plays as lead agency for matters relating to this arrangement, and in the emerging regime for the Arctic where a dozen or more agencies demand a say in what happens but none is able or willing to accept the role of lead agency (Osherenko and Young 1989, chapter 8).

As this discussion makes clear, international regimes normally operate in social settings featuring substantial institutional heterogeneity among their members. What is more, those responsible for administering the regimes are seldom in a position to resort to what constitutes the normal procedure for handling interplay between national and subnational governments. In this setting, national governments generally have ultimate authority to compel subnational governments to adjust their rules and procedures to ensure that they do not conflict with arrangements established at the national level. The result is a system in which rules of international regimes are framed in terms that are sufficiently generic to allow officials in member states considerable leeway in operationalizing them within their own jurisdictions. Up to a point, this is clearly desirable. National officials are not about to let the managers of international regimes dictate to them, and much can be said for allowing individual members to assimilate the rules of international regimes into their own systems in ways they deem appropriate. But this accentuates the argument under consideration here to the effect that the consequences of international regimes will be determined in considerable part by the interplay between the regimes themselves and national practices prevailing in member states. Among other things, this should lead us to expect considerable variance in the performance of member states when it comes to fulfilling commitments made during regime formation. This variance may not be critical to the overall performance of international regimes. In the case of equipment standards applicable to construction of oil tankers, for instance, the regime can be expected to operate effectively as long as a few key member states take the standards seriously (Mitchell 1994). But in other cases, such as phasing out production and consumption of ozone-depleting chemicals (French 1997), it is apparent that it takes conformance on the part of all but the most marginal members to achieve effective protection of relevant natural systems.

Regimes for Terrestrial Resources

To think concretely about the impact of this form of interplay on patterns of land use, consider some examples relating to the operation of the International Tropical Timber Agreement (ITTA) and the effort to create a Circumpolar Protected Areas Network (CPAN) in the Far North. The ITTA, created initially in 1983 and substantially restructured in 1994, is first and foremost a trade agreement in which producers (e.g., Indonesia, Malaysia, and the Philippines) and consumers (e.g., Japan) of tropical timber endeavor to stabilize and regulate the world market in wood products harvested from tropical forests (Humphreys 1996; Dauvergne 1997b). What makes this regime interesting from an environmental point of view is the recognition that a high proportion of harvesting of tropical timber in recent decades has taken the form of destructive practices best described as the "mining" of forests and that it is necessary to restructure the industry to put it on a more sustainable basis. The centerpiece of the 1994 agreement is a commitment on the part of regime members to implement guidelines to ensure that both natural and planted tropical forests are managed sustainably and that biological diversity is protected. To this end, the members committed themselves to the Year 2000 Objective calling for all tropical timber entering international trade to be produced from forests under sustainable management by the year 2000. Only a few countries succeeded in meeting this goal.[12] Are others likely to be able to meet it in the near future? The answer depends on interplay between the international regime and the national political systems of member countries, such as Indonesia and Japan (Guppy 1996). At this stage, the prognosis is not encouraging. Given the economic and political turmoil affecting southeast Asia combined with the continuing grip of crony capitalism, the capacity of a country such as Indonesia to reach the Year 2000 Objective is limited, and the sanctions associated with nonconformance are likely to prove ineffectual. For its part, the severity of the economic downturn that plagues Japan, together with the political influence of the major companies involved in the tropical timber trade, creates a setting that is not conducive to bringing effective pressure to bear on domestic users of tropical timber.

A major goal of the Arctic Environmental Protection Strategy (AEPS), launched in 1991 but integrated since 1996 into the broader structure of the Arctic Council, is to promote conservation of flora and fauna in the circumpolar North (Huntington 1997). To this end, the AEPS established a Working Group on the Conservation of Arctic Flora and Fauna (CAFF) and provided it with a mandate to take the initiative in devising innovative means to achieve its general goal. Despite the relative weakness of CAFF in terms of formal authority, this initiative has generated a good deal of interest. Indeed, CAFF has become a forum in which government officials and representatives of nonstate actors interact freely; it has succeeded in capturing and holding the attention of public agencies in a number of member states; and it has emerged as a mechanism for applying universal guidelines relating to biological diversity to the particular circumstances prevailing in the circumpolar North.[13] One of CAFF's highest priorities has been to promote and oversee the creation of CPAN, a linked system of parks, preserves, wildlife refuges, and so forth located in all the Arctic countries and organized in such a way as to provide harmonized management for the entire system (CAFF 1996). The success of this initiative depends largely on the willingness and ability of management agencies in member states to collaborate effectively and manage protected natural areas on a coordinated basis.

This is where problems begin to arise in connection with this intuitively appealing initiative. In some key countries—the United States is a good example—management authority regarding the areas involved resides with a number of distinct agencies (e.g., National Parks Service, Fish and Wildlife Service, Geological Survey, Bureau of Land Management) that are not in the habit of cooperating effectively with one another, much less with their counterparts in other countries (Clarke and McCool 1996). In other countries—the Russian Federation, for example—economic and political problems are so severe that little energy and few resources are available for international cooperation. This initiative does not require integrated management across national jurisdictional boundaries; coordinated or harmonized management practices carried out by relevant agencies in each country would suffice. Yet the complexities of institutional interplay between international commitments and national practices raise serious questions about the prospects for CPAN.

Regimes for Marine Resources
Turning to institutional interplay relating to marine resources in the Barents Sea and the Bering Sea, an even more complex pattern comes into focus. In effect, regimes that emerged in these areas feature interactions

between and among three distinct sets of institutional arrangements: global rules governing EEZs, national and subnational regulatory systems that individual coastal states have put in place in their own EEZs, and several regional arrangements created to deal with situations in which national EEZs either adjoin each other (i.e., relevant states are adjacent or opposite states) or leave pockets of high seas surrounded by national EEZs. Although the introduction of EEZs was justified in large measure as an innovation required to manage human uses of marine living resources on a sustainable basis, it soon became apparent that it created a range of new problems, quite apart from its consequences with regard to treatment of preexisting problems. Marine ecosystems do not conform to legal or political boundaries, however ingenious the effort to delineate them may be. As a result, many states that acquired expanded jurisdiction over harvesting of living resources in their individual EEZs now are confronted with a sizable collection of new problems relating to what are known as straddling stocks. One response to this development, intended to coordinate efforts to manage marine resources located partly within an EEZ and partly in the high seas, is embodied in the Straddling Fish Stocks Agreement, a global arrangement negotiated in the wake of the U.N. Conference on Environment and Development and signed in 1995.[14] Another response, intended primarily to coordinate the efforts of adjacent and opposite states to manage fish stocks common to their individual EEZs and related areas of the high seas, takes the form of a growing collection of regional fisheries regimes.

Two particularly interesting examples of these regional arrangements are the predominantly bilateral Norwegian-Russian regime dealing with the fisheries of the Barents Sea and the multilateral arrangements that have emerged in the Bering Sea region (Stokke 2001a). Not only do they exemplify different ways of dealing with institutional interplay, they have produced significantly different outcomes. In the Barents Sea, Norway and Russia capitalized on EEZs during the 1970s to create a bilateral regime that phased out or drastically curtailed participation on the part of fishers from third states and put in place a system under which the principal fish stocks of the entire region are managed on an integrated basis (Stokke, Anderson, and Mirovitskaya 1999). This system is not immune to biogeophysical surprises. It has had to adjust to changing biological conditions (e.g., location of spring spawning herring), and it has had to cope with severe stresses attributable to the transition from the Soviet Union to the Russian Federation and subsequent decline in the capacity of Russia to regulate the activities of Russian fishers (Hønneland 2000; Stokke 2001b). But this is a case in which interplay between two sets of national arrangements and an international regime has been managed in such a way as to avoid crises requiring the closure of important fisheries.

In contrast, the situation that emerged in the Bering Sea region illustrates a somewhat less auspicious response to institutional interplay. Russia and the United States responded to the creation of EEZs by establishing complex but somewhat poorly coordinated national regimes in the western and eastern Bering Sea areas, respectively. In addition, the 1990s brought creation of a regional agreement covering salmon stocks migrating back and forth through the EEZs of the two countries, together with a six-nation agreement dealing with pollock stocks of the doughnut hole and designed to prevent a recurrence of the collapse of these stocks that occurred in the late 1980s and early 1990s. But the results of this complex mosaic are far from reassuring. Both coastal states experienced problems controlling harvests of living marine resources within their own EEZs. The pollock stocks of the doughnut hole have not recovered sufficiently to allow for harvesting under the terms of the international agreement. Above all, a number of disturbing indications suggest that anthropogenic forces have triggered severe stresses affecting the Bering Sea ecosystem as a whole (National Research Council 1996; National Marine Fisheries Service 1997; WWF/TNC 1999). These include startling declines in populations of several unharvested species, such as sea lions, northern fur seals, sea otters, and red-legged kittiwakes, as well as some harvested species, such as spectacled eiders and several species of geese. No doubt, it would be wrong to point to problems of institutional interplay as the sole or even primary cause of these disturbing developments. But it is hard to avoid the conclusion that difficulties plaguing efforts to coordinate institutional arrangements across levels of social organization are a significant feature of this story.

Implications and Take-Home Messages

The principal conclusion to be drawn from the analysis presented in this chapter is that cross-scale interactions among resource regimes generate inescapable tension between the benefits and costs of higher-level arrangements. The move to higher levels of social organization makes it possible to consider interdependencies in large marine and terrestrial ecosystems and to devise regimes based on the precepts of ecosystem management. Yet costs to operating at higher levels are measured in terms of inability to come to terms with local variations in biogeophysical conditions and lack of sensitivity to both the knowledge and the rights and interests of local stakeholders.

Those operating at higher—national or international—levels are typically compelled to devise and promulgate structures of rights and regulatory rules in terms that are broad and generic. Whereas this may cause few problems in dealing with large marine and terrestrial ecosystems that are homogeneous, problems mount rapidly when local variations occur both in pertinent biogeophysical conditions (e.g., population dynamics of fish stocks) and in patterns of human uses of natural resources (e.g., hunting and herding practices). The result more often than not is proliferation of formal rights and rules that are poorly suited to local circumstances or evolution of systems so encrusted with local exceptions and informal interpretations that they become unworkable.

The same is true regarding the rights and interests of various groups of stakeholders. Moving to higher levels of social organization can open up opportunities for increased efficiency in the use of resources and for more comprehensive approaches to equity, but the associated costs are often substantial. National regimes increase the influence of economically and politically powerful actors who do not reside within the ecosystems they exploit, who move on to new areas once the resources of one area are exhausted, and who favor exploitation of resources that are tradable in (often international) markets. For their part, international regimes often cater to the interests of multinational corporations that have operations in many places and that have no long-term commitment to the ecological welfare of particular areas or the social welfare of those who reside permanently in these areas. Under the circumstances, it is easy to see that shifts to higher levels of social organization, justified to manage large marine and terrestrial ecosystems in a holistic manner, often lead to changes in patterns of land and sea use that raise profound questions not only in terms of sustainability but also in terms of normative concerns such as equity and efficiency.

The vigor of the debate about what is often called the subsidiarity principle is testimony to the importance of this tension regarding environmental consequences of cross-scale interactions.[15] But this principle, which calls for creation of management systems that lodge authority at the lowest level of social organization capable of solving pertinent problems, does not offer much help in coming to terms with problems of cross-scale interplay. National and even international arrangements are necessary to manage human activities relating to large marine and terrestrial ecosystems. Yet the dangers inherent in moving from local to national and from national to international regimes are severe. What is needed under the circumstances is a conscious effort to design and operate institutional arrangements that take local knowledge seriously and protect the rights and interests of local stakeholders, even while they introduce mechanisms at higher levels of social organization required to encompass the dynamics of ecosystems that are regional and even global in scope.

This is not a task to be handled by unequivocal decisions regarding the proper level of social organization at which to vest management authority. A more interesting response involves arrangements that numerous analysts have explored under the rubric of comanagement (Osherenko 1988; Berkes 2000). Typically, comanagement involves creation of an environmental or resource regime featuring working partnerships between local users of natural resources and (sub)national agencies with the formal authority to make decisions about human activities involving marine and terrestrial ecosystems as well as the resources to administer management systems. This intrinsically appealing approach may well give rise to a range of social practices that are of lasting significance in dealing with problems of vertical interplay; but it would be premature to jump to any such conclusion at this stage. Comanagement is in danger of becoming a catch-all conceptual category containing a ragtag collection of tenuously related arrangements. Even in dealing with interplay between local and national arrangements, experience on the ground is limited, and we are far from formulating well-tested propositions about the determinants of success and failure in the creation and operation of comanagement regimes. It is anything but clear whether experience with comanagement in dealing with local-national interactions can be scaled up to offer an effective method of organizing the interplay between national and international regimes. These observations are not meant to belittle the significance of a strategy involving political design and management as a means of coming to terms with problems arising from functional interdependencies; many analysts are engaged in interesting studies of comanagement at the present time. Nonetheless, much remains to be done before we can assert that substantial progress is being made in coming to terms with tensions arising from cross-scale interactions.

We must bear in mind as well that the creation of institutions at every level of social organization is a political process centering on what is often called institutional bargaining (Young 1994a). Whatever their consequences in terms of considerations such as sustainability and efficiency, environmental or resource regimes always have significant consequences for the interests of those—nonstate actors as well as states—subject to their rules and decision-making procedures. It should come as no surprise, therefore, that individual actors often work hard to advance their own causes during regime formation, and outcomes typically reflect the political influence of major participants or coalitions of participants. This is not to suggest that efforts to design institutions that will promote social goals such as sustainability or efficiency will always be exercises in futility. In fact, institutional bargaining has some features that make it more open to design considerations than conventional or distributive bargaining. Yet we cannot escape the fact that regime formation is better understood as a political process in which bargaining strength plays a central role than as an exercise in social engineering in which apolitical design principles predominate.

Conclusion

This chapter is intended to initiate a dialogue regarding the role that interplay involving cross-scale interactions among distinct institutions plays in the overarching picture of the human dimensions of environmental change. Land use and sea use are particularly interesting in this connection because their patterns are directly and intimately linked to large-scale environmental changes such as loss of biological diversity and climate change. But similar issues of institutional interplay arise in conjunction with other concerns, including human uses of atmospheric and hydrological systems. It cannot be assumed that institutions in general or the interplay among distinct institutions in particular can account for all the variance in the effects of human actions on atmospheric, hydrological, marine, or terrestrial systems. On the contrary, institutional drivers interact with other forces in complex ways; one of the main challenges facing those interested in the human dimensions of large-scale environmental change is to sort out the relative significance or weight of institutional drivers and other driving forces.

Yet emphasis on the role of institutions in this connection has great appeal as long as care is taken to avoid the assumption that institutional arrangements operate in a vacuum, in the sense that they produce results without regard to the character of the broader biogeophysical and socioeconomic settings in which they operate. The content of prevailing institutions is subject to intentional reform, a fact that opens up the opportunity to respond to functional interdependencies by turning to political design and management in the interests of minimizing negative consequences of existing institutions and supplementing or even replacing these arrangements to mitigate or adapt to large-scale environmental changes. The message of this chapter regarding this prospect is one of great caution but certainly not one of pessimism. Even if we succeed in identifying institutional forces giving rise to environment problems, we have no guarantee that we can take steps to alter the operation of prevailing arrangements in a well-planned and efficacious fashion. Nonetheless, the prospect that (re)designing institutions can play a role in controlling or managing large-scale environmental changes provides a compelling reason to invest time and energy in enhancing our understanding of the dynamics of institutional interplay.

5

Horizontal Interplay: The Politics of Institutional Linkages

Although those seeking to understand the formation and operation of individual regimes may be tempted to deemphasize or even ignore events occurring beyond the confines of their own cases, linkages among institutions operating at the same level of social organization are ubiquitous. Sometimes these links arise from functional interdependencies among the activities covered by different regimes (e.g., separate arrangements dealing with commercial and environmental matters). Other horizontal interactions are attributable to the fact that individual regimes are based on premises or principles that are cross-cutting or orthogonal to one another (e.g., the global but functionally restricted regime for whales and the geographically limited but functionally broader regime covering Antarctic marine living resources). In still other cases, horizontal interplay results from deliberate efforts of individual actors or interest groups to pursue their own objectives by developing competing regimes regarding a single issue.

More generally, it is fair to say that the extent and significance of horizontal interplay are functions of the density of institutional arrangements operative in a society. As the number of institutions in a given social space rises, opportunities for interactions between and among individual arrangements increase exponentially. This explains why horizontal interplay has long been a subject of great interest to those who focus on (sub)-national systems. It also accounts both for the lower-level of interest in such matters in the past among those concerned with international relations and for the growing interest among analysts concerned with international society.

Many horizontal linkages, much like vertical linkages, take the form of side effects or unintended byproducts of actions designed to achieve other ends. A number of analysts speak of institutional overlaps to refer to linkages of this sort (Herr 1995). Even so, players active in most social settings are acutely aware of the fact that the regimes they create are likely to interact with other institutional arrangements at the same level of social organization; many incorporate this awareness into their goal-directed behavior as a matter of course. In some cases, responses to this awareness are guided predominantly by a desire to solve problems or enhance cooperative outcomes, and as a result, to make distinct institutional arrangements fit together into structures that promote the common good. In other cases, those concerned with horizontal interplay strive to exploit linkages among institutional arrangements for competitive purposes, to advance their own agendas whether or not such actions detract from joint initiatives (Schelling 1960). Taken together, conscious efforts to make use of interplay to promote both cooperative and competitive ends constitute a domain of activities that can be thought of as the politics of institutional linkages.

This chapter addresses this topic in stages. The first two sections focus on efforts to structure institutional interplay primarily through exercises in joint decision making intended to maximize or at least enhance social welfare. The analysis turns first to formative links or institutional interactions arising during regime (re)formation and then to operational links or institutional interactions arising in connection with day-to-day operation of regimes once they are in place. The final section explores strategic uses of institutional interplay in which influential actors endeavor to exploit links among distinct arrangements to advance their own causes. As in other chapters, this account directs attention to examples involving the institutional dimensions of large-scale environmental changes.

The general conclusion arising from this analysis is straightforward. Institutional interplay ordinarily generates incentives to manage interactions in such a way as to reap joint gains or avoid joint losses. But achieving this goal is easier said than done. Not only does interplay involve mixed-motive situations in which actors often act in ways that complicate pursuit of the common good (Schelling 1960); the resultant interactions sometimes open up opportunities for strategic behavior on the part of those who have little or no interest in promoting the common good. This is not to say that the politics of institutional design and management always lead to failure from the perspective of maximizing social welfare. But it does provide grounds for proceeding with caution and avoiding naïve expectations in this realm.

Formative Links

Those who take the initiative in efforts to form institutions as a means of responding to newly emerging problems, or to reform existing institutions in order to adapt them to changing circumstances, must make a number of decisions that have implications for institutional interplay. They may address a broad range of more or less related issues within the ambit of a comprehensive institutional arrangement, an approach that will yield clustered regimes (e.g., comprehensive arrangements articulated in the 1982 U.N. Convention on the Law of the Sea) in which linkages among different regimes are minimized but the internal complexity of the resultant arrangements rises rapidly. Alternatively, they may define problems narrowly and endeavor to create distinct regimes focusing on a variety of separate issues (e.g., conservation of migratory birds, protection of marine mammals, management of fish stocks). In such cases, the scope for interplay among regimes increases proportionately. No doubt, in some instances choices of this sort are made in an unreflective manner so that subsequent interactions are properly thought of as institutional overlaps. More often than not, however, actors take such decisions self-consciously, which gives rise to the politics of institutional linkages, whether or not the ultimate outcomes conform to the expectations of those whose arguments carry the day during processes of regime (re)formation.

Framing the Issues

Regimes are problem driven. Yet framing issues to be addressed by specific regimes is a social process that is not determined entirely by objective characteristics of the relevant problems (Litfin 1994; Wendt 1999). Consider climate change. Even when it is addressed in isolation from other atmospheric concerns (e.g., ozone depletion, acid precipitation), there are a number of different ways to conceptualize the issue. Is it better to concentrate on carbon dioxide on the grounds that this is the most important greenhouse gas, or to think in terms of a basket of gases on the grounds that the problem cannot be solved without controlling atmospheric concentrations of the full range of greenhouse gases? Whatever the range of gases included, does it make sense to focus exclusively on arrangements dealing with reductions in emissions, or to consider arrangements that direct attention to combinations of emissions reductions and emissions offsets (in such forms as the sequestration of carbon dioxide in forest sinks)? Is it preferable to separate the problem of reducing current emissions on the part of advanced industrial countries from the problem of avoiding increases in future emissions on the part of developing countries, or conversely, to opt for integrated arrangements featuring joint implementation and a clean development mechanism intended to encourage both advanced industrial and developing countries to enter into mutually beneficial relationships in this area?

Beyond this lies the question of the relative merits of focusing on climate change as an issue to be addressed on its own terms (without reference to other atmospheric concerns) or recognizing explicitly the links between this and other atmospheric problems, such as depletion of stratospheric ozone and long-range transport of airborne pollutants. The functional interdependencies between these problems are well known. A number of proposed substitutes for chlorofluorocarbons (CFCs) are themselves greenhouse gases; the presence of sulfur dioxide in the atmosphere is likely to retard or mitigate global warming. Other links are poorly understood at present, but no less relevant. Some evidence, for instance, suggests links between emission of greenhouse gases and ozone depletion. The basic idea is that warming of the troposphere associated with the greenhouse effect may lead to cooling of the stratosphere and a resultant increase in the rate of ozone depletion. Responding to these concerns, some observers have suggested that serious thought be given to developing a law of the atmosphere; that is, a clustered regime covering a range of anthropogenic effects on atmospheric systems in much the same way that the law of the sea encompasses a range of issues dealing with the effects of human activities on marine systems.

Treatment of these atmospheric issues also poses questions involving links between environmental regimes and economic arrangements pertaining to matters such as international trade. The ozone regime authorizes trade sanctions to induce or compel actors to fulfill commitments regarding production and consumption of ozone-depleting substances. Effective measures to combat climate change will require restrictions on production processes for energy-intensive goods that enter into international trade. Many suggestions for reducing emissions of greenhouse gases or encouraging investments in emissions offsets envision establishment of quasi-markets that would be subject to the general rules governing international trade. Under the circumstances, it is impossible to avoid a consideration of the relative merits of merging arrangements covering trade and climate change on the one hand, or dealing explicitly with the interplay between separate arrangements covering these matters on the other.

As this illustration makes clear, those endeavoring to frame issues to set the stage for efforts to (re)form regimes often confront options ranging from the widest possible formulation in which the full array of issues relating to a broad area of human activity are treated as the appropriate domain, to highly restrictive formulations in which issues are defined as narrowly as possible and addressed separately in efforts to create effective institutions.[1] The choices that actors make about such matters are neither right nor wrong in any objective sense. But they do have strikingly different consequences with regard to horizontal interplay. As the scope of issues encompassed by a single institutional arrangement expands, opportunities for interplay with separate regimes decline but internal complexities associated with the operation of the arrangement grow and vice versa. It is no surprise that advantages and disadvantages are associated with efforts both to narrow and to expand the scope of specific regimes. The inner workings of narrow regimes are relatively simple, but they typically require greater investments of time and energy to manage interactions with other regimes. Conversely, more comprehensive regimes are able to endogenize institutional linkages, but they are likely to require much greater attention to issues arising from internal complexities or contradictions and the rigidities or potential for stalemate that go with them.

In general terms, the optimum in this context occurs at the point at which marginal costs arising from increases in institutional interactions just equal marginal costs associated with increases in internal complexity. Yet the value of a general formula of this sort is largely heuristic. It is improbable that relevant benefits and costs arising in specific situations can be measured with sufficient precision to yield conclusions about the optimum that will prove useful for purposes of policy making. How, then, are such decisions about alternative approaches to framing the issues made in actual instances of regime (re)formation? I maintain that three distinct, although by no means mutually exclusive, factors account for most of the variance in this area: cognitive fashions, organizational mandates, and actor interests.

Fashions, in the form of demonstration effects or simple extrapolations from recent experiences, are common in this realm. For instance, many practitioners have taken note of difficulties encountered in the effort to reform the law of the sea during the 1970s and 1980s and attributed these difficulties to the decision to create a comprehensive arrangement. They typically conclude that it is better to approach large-scale environmental issues in relatively narrow or disaggregated terms (Susskind 1994). As a result, efforts to form regimes during the 1980s and 1990s often focused on issues defined in relatively narrow terms, such as ozone depletion, acid precipitation, or transboundary movements of hazardous wastes. The apparent success of the regime dealing with CFCs and other chemicals implicated in the depletion of stratospheric ozone reinforced this reaction to complications expected to afflict efforts to devise a clustered regime covering a wide range of human activities affecting atmospheric systems (Soroos 1997). Yet this trend is an ad hoc response to experience rather than an approach to handling large-scale environmental problems that is well grounded in a comparative assessment of the costs and benefits of alternative strategies. Specific problems such as ozone depletion and climate change are linked in functional terms, which ensures that a good deal of interplay will occur among related but separate regimes, whatever the preferences of those who negotiate the terms of individual arrangements. It remains to be seen whether and when the costs of endogenizing links of this kind through creation of more comprehensive regimes exceed the costs of dealing with unavoidable interactions among separate and more narrowly defined arrangements.

In other cases, organizational mandates influence or even dictate the manner in which issues are framed for regime formation. Many observers have pointed out, for example, that marine pollution is attributable to a combination of vessel-source discharges and land-based runoffs. Evidence is growing that land-based runoffs are a critical concern, a conclusion that suggests that efforts to come to terms with pollution in large marine ecosystems should encompass both vessel-source and land-based pollutants in a coordinated fashion. Yet as long as marine pollution is handled through agencies that deal exclusively with vessel-source pollution, such as the International Maritime Organization (IMO), there is little prospect that a coordinated approach to the problem will be adopted. Dealing with vessel-source problems is a central concern of the IMO; coping with land-based problems is not part of its mandate (Mitchell 1994). This situation undoubtedly helps to account for the contrast between the scope of the International Convention for the Prevention of Pollution from Ships (MARPOL 1973–1978) which is limited to vessel-source pollution, and the scope of the regime for the North Sea, which was developed during the same time period but covers land-based pollutants as well as dumping at sea. Whereas MARPOL is associated with the IMO, the North Sea regime was created as a free-standing arrangement under the terms of the Oslo and Paris Conventions (Mitchell 1994; Skjærseth 2000).[2] These considerations have little to do with any generic test for optimality in determining the balance between endogenization and interplay. But they do help to explain patterns of institutional interactions that occur in specific situations.

Nor should we ignore the interests of key actors as determinants of the framing of issues that have significant implications for institutional interplay. With regard to climate change, the United States and other members of the Umbrella Group have consistently favored an approach encompassing the full set of greenhouse gases. Members of the European Union, however, have often argued for a more restrictive approach with respect to the gases covered. The interests underlying these divergent preferences are not hard to identify. Members of the Umbrella Group assume that they will find it easier to fulfill commitments relating to targets and timetables if they have more room to maneuver and, arguably, to find ways to meet the letter of these commitments while avoiding painful changes in current practices. A particularly transparent example is the view that actions associated with phasing out CFCs and other ozone-depleting substances should be counted toward fulfilling commitments to reduce net emissions of greenhouse gases. For their part, members of the European Union have a strong interest in finding ways to decarbonize their economies; they regard emissions of carbon dioxide as the main feature of climate change. Taking the problem seriously therefore requires an approach that compels those subject to provisions of the climate regime to come to terms with complications associated with meaningful efforts to reduce their carbon emissions. The point is not that one or the other of these approaches to climate change is correct. But the two approaches can be expected to have quite different implications for the balance between endogenization and institutional interplay. These differences clearly involve the politics of institutional linkages in the sense that they give rise to intentional acts. But there is no basis for assuming that resultant interactions will produce optimal results measured in terms of institutional arrangements that contribute to solving the problem of climate change.

Choosing Arenas
Choices pertaining to arenas in which regime formation takes place often go hand in hand with framing issues for purposes of creating institutional arrangements. Sometimes framing actually dictates the choice of arenas and vice versa. Yet it is worth drawing a distinction between framing issues and choosing arenas, and considering how the latter can affect the scope for and nature of institutional interplay. Nothing is self-evident or automatic about the choice of arenas for regime formation. The Intergovernmental Negotiating Committee on Climate Change (INC), the body responsible for negotiating the terms of the 1992 U.N. Framework Convention on Climate Change, operated directly under the auspices of the U.N. General Assembly despite the fact that many participants would have preferred a process managed by the U.N. Environment Programme (UNEP) or one involving a more central role for the Intergovernmental

Panel on Climate Change (IPCC). In contrast, the Intergovernmental Negotiating Committee on Biological Diversity, an entity created at roughly the same time and given the assignment of devising the provisions of the 1992 Convention on Biological Diversity, operated under the auspices of UNEP.

Nor are these isolated instances. Those desiring to negotiate the terms of what became the 1979 Convention on Long-Range Transboundary Air Pollution (CLRTAP) confronted a choice of negotiating forums including the Council of Europe, the European Union (then the European Economic Community), the Organization for Economic Cooperation and Development, and the U.N. Economic Commission for Europe (UNECE). Ultimately, they selected the UNECE as the appropriate vehicle for this endeavor, a decision that has had far-reaching consequences for the evolution of this regime over the last twenty years. The 1995 Straddling Fish Stocks Agreement emerged from a series of negotiating sessions taking place under the auspices of the U.N. General Assembly. The Code of Conduct for Responsible Fisheries, negotiated at the same time and also made final in 1995, was the product of a process taking place under the auspices of the U.N. Food and Agriculture Organization (FAO), one of a collection of specialized agencies that make up the United Nations System.[3] Yet another procedure is exemplified by free-standing processes leading to creation of the whaling regime through the adoption of the 1946 International Convention on the Regulation of Whaling and formation of the regime for Antarctica through adoption of the 1959 Antarctic Treaty. In the case of Antarctica, the parties have made a point of avoiding arenas linked to the United Nations and actively resisting efforts to bring this regime into some recognized relationship with the United Nations (Joyner 1998).

What consequences do these choices regarding arenas have for patterns of institutional interplay that grow up around regimes? This question has no simple answers. But it is apparent that the consequences can be far-reaching. The decision to create a negotiating committee for climate change reporting directly to the U.N. General Assembly signaled both that the North-South dimensions of this problem would figure prominently in regime creation and that climate change would become a matter of "high" politics rather than one regarded as suitable for largely scientific or exclusively environmental consideration. This choice stimulated emergence of institutional interplay featuring prominent interactions between climate change and the larger economic arrangements associated with globalization, as well as broader political arrangements governing North-South relations. In the case of CLRTAP, the choice of the UNECE, in contrast to the European Union or the OECD, as a negotiating forum reflected both the significance of this issue in East-West terms and a desire to avoid thinking about the problem as a purely European matter. This decision ensured the emergence of institutional interplay featuring links between air pollution in Europe and North America and, more broadly, between this environmental concern and various efforts to alleviate tensions of the Cold War.

Similar remarks are in order about other cases just referred to. The negotiating process that produced the 1995 Straddling Fish Stocks Agreement led to close association between this arrangement and the overarching institutional framework established under the terms of the U.N. Convention on the Law of the Sea (UNCLOS). For better or for worse, the regime for fish stocks will interact with other international regimes as a part of this nested system rather than as an arrangement with a life of its own. The origins of the regimes for whaling and Antarctica, on the other hand, have produced the opposite result. Needless to say, extensive interplay exists between these regimes and other institutional arrangements operative at the international level. The whaling regime, which is global in scope but functionally narrow, and the regime for Antarctica, which is functionally broader but geographically limited, even exhibit functional interdependencies with one another. Given their origins, however, it is difficult to come to terms with institutional interplay in this case through the organizational structures provided by the United Nations System.

For the most part, choices of arenas for regime formation involve conscious acts, so the resulting institutional interactions deserve to be included in consideration of the politics of institutional linkages. But one cannot conclude from this observation that the choices are products of sophisticated efforts to design arrangements that will prove optimal in terms of problem solving. Rather, the outcomes are better understood as products of organizational imperatives on the one hand and calculations of actor interests on the other. There is no mystery about the desire of UNEP to control regime formation in the area of climate change, of the FAO to dominate the creation of fisheries arrangements, or of the IMO to influence the content of regimes dealing with marine pollution.[4] The mandates of these organizations cover the issues in question and provide powerful incentives to seek to control the action when it comes to regime building in their areas of competence. Failure to do so inevitably raises questions about the authority of these organizations in their own fields and leads to erosion of their ability to influence the course of events in international society.

Yet it is equally easy to understand why major players or coalitions engaged in regime formation are often unwilling to allow such processes to proceed in the arenas provided by these organizations. Developed countries apparently distrusted the leadership of Mostafa Tolba, UNEP's Executive Director at the time, in the case of climate change (Tolba 1998). For their part, some developing countries were not confident that a process dominated by UNEP would produce a climate regime responsive to their concerns. Those motivated by concern for conservation were not interested in the more comprehensive approach to fisheries issues advocated by FAO and lacked confidence that FAO would be able to produce arrangements adequate to avoid severe depletions of straddling stocks and highly migratory species. The great powers were determined to handle the problem of Antarctica in an arena of their own devising that they would be able to control without having to contend with the interests of a large number of (in their view) extraneous concerns introduced by members of the U.N. General Assembly. The UNECE became the arena of choice for dealing with long-range transboundary air pollution not because of established competence in handling environmental matters, but because it seemed to offer a setting congenial to those desiring to use this regime as a means of alleviating broader East-West tensions. The ultimate outcomes, stemming as they do from the confrontation of organizational imperatives and interests of major players in the negotiating processes, are unlikely to yield results that are optimal in terms of institutional interplay. Yet these formative links can and often do set the stage for the emergence of patterns of interplay whose effects are felt long after regime formation is complete.

Bargaining over Content

Framing an issue and selecting an arena in which to conduct negotiations about it set the stage for institutional bargaining; that is, bargaining over the contents of constitutional contracts—conventions, treaties, declarations, or other explicit agreements—in which provisions of regimes are articulated (Young 1994a). Many observers have noted the role that adding or subtracting issues plays in shaping the course of institutional bargaining and, in the process, determining the scope of the politics of institutional linkages (Sebenius 1983). Broadly speaking, adding issues as a means of facilitating institutional bargaining will have the effect of endogenizing linkages by expanding the range of concerns in individual regimes. Subtraction of issues can be expected to have the opposite effect.

Once again, climate change illustrates these dynamics. The bargaining process began with a focus on reducing emissions of greenhouse gases and especially carbon dioxide. In an effort to make the emerging regime palatable to key countries (e.g., the United States), however, the options of meeting targets and timetables through mechanisms such as emissions trading and joint implementation and of lowering net emissions by making use of emissions offsets (in such forms as the uptake of carbon by aforestation) were added as integral parts of the arrangement. Similarly, provisions dealing with creation of a clean development mechanism and additionality with regard to funding were introduced to make the regime more attractive to developing countries whose emissions of greenhouse gases are currently modest but that are on a course that will make some of them major emitters within the next three to five decades.

Institutional bargaining produced a steady expansion in the scope of the regime and, as a result, a significant trend toward endogenization of institutional linkages. Even so, the process has limits. Whereas functional interdependencies between control of greenhouse gases and regulation of international trade are likely to be substantial, no one has launched a serious effort to integrate the regime for climate change and arrangements governing international trade articulated in the agreement establishing the World Trade Organization (WTO). In more general terms, however, these observations prompt an inquiry into the distinguishing features of institutional bargaining at the international level and their consequences for the politics of institutional linkages.

Bargaining over provisions of regimes dealing with large-scale environmental problems generally involves many actors who strive to devise formulas acceptable to all participants or, at least, to all the major players. In cases involving deep-seated problems, complexities arising from the participation of large numbers of states—more than 180 in the case of climate change—are commonly alleviated by formation of a small number of coalitions or blocs that become the main protagonists in the bargaining process. Negotiations relating to climate change are now dominated by four coalitions: the Umbrella Group, the European Union, the Group of 77 plus China, and the Alliance of Small Island States. Institutional bargaining generally is a mixed-motive process featuring a substantial component of what is often described as integrative or productive bargaining, as well as a significant element of distributive or positional bargaining (Schelling 1960; Walton and McKersie 1965). The integrative component arises from the facts that information is imperfect, the locus of the welfare frontier is unknown or poorly delimited, and the veil of uncertainty is thick enough to provide participants with incentives to design mutually beneficial arrangements (Brennan and Buchanan 1985). The balance between these integrative forces and concomitant incentives to focus on distributive concerns varies from one instance of institutional bargaining to another. But for the most part, those seeking to arrive at mutually acceptable formulas under such conditions proceed by developing and refining negotiating texts that become vehicles for moving the process from articulation of a broad range of disparate perspectives to formulation of specific provisions that all major coalitions are willing to accept as components of a constitutional contract.

What are the implications of these features of institutional bargaining for the politics of institutional linkages? The emphasis on consensus building and the open character of the process arising from the search for new approaches under conditions of imperfect information normally expand the scope of arrangements under consideration. As long as it is deemed essential to satisfy all the major players or coalitions, participants will experience incentives to add new elements in hope of putting together packages that contain enough benefits for all concerned to provide them with convincing reasons to accept the terms of the final agreement (i.e., sign and, if necessary, ratify the convention or declaration). The fact that this sort of bargaining is not simply a matter of "life on the Pareto frontier" (Krasner 1991) reinforces this tendency to add issues as a means of satisfying the principal concerns of all major parties. Institutional bargaining is seldom dominated by threats and committal tactics to compel others to accept particular outcomes on well-defined welfare frontiers or contract curves. There is nothing to prevent the parties from taking steps to incorporate new issues if that appears to be the most effective way to break deadlocks in the effort to arrive at generally acceptable formulas.

Does this mean that we should expect institutional bargaining, at least at the international level, to produce comprehensive or clustered regimes and thus reduce the scope for interactions with other distinct regimes? Not necessarily. Bargaining of this sort seldom leads to regimes that cut across the boundaries of broad issue areas, if only because that would require integrating the efforts of agencies in the governments of member states that are not in the habit of working together. It is unlikely, under the circumstances, that a climate regime that integrates provisions covering purely environmental matters with provisions relating to international trade and financial flows will emerge, despite the fact that there is much to recommend such an approach from the point of view of problem solving. Arguably, the environmental side agreement grafted onto the North American Free Trade Agreement is an exception. But this is a highly unusual arrangement, and many observers see the side agreement as a kind of afterthought introduced to make the overall agreement more palatable in political terms, rather than as a serious example of integrated regime building. Beyond this, using negotiating texts to reach closure in institutional bargaining has the effect of narrowing the scope of the formulas ultimately adopted. The emphasis is on a search for common denominators that leads to the funneling of a wide range of initial suggestions into a smaller and smaller stream of provisions formulated in such a way as to facilitate their codification as elements of an international convention or treaty.[5] As the United States found out in 1981 when it sought to reopen carefully crafted agreements in the revised single negotiating text for the U.N. Convention on the Law of the Sea, attempting to revisit major issues in a way that casts doubt on earlier agreements is an unpopular step that even a superpower may be unable to take without incurring substantial costs (Sebenius 1984; Friedheim 1993).

As in the cases of framing issues and choosing arenas, bargaining over the content of constitutional contracts can have far-reaching consequences for the politics of institutional linkages. But here, too, one cannot necessarily expect the process to be guided by a search for an optimal balance between endogenization and institutional interplay. Institutional bargaining is driven by pressures to craft packages of arrangements that prove acceptable to major players or coalitions operating under competitive-cooperative conditions in which information is far from perfect. The results reflect the limits of political feasibility and may involve substantial elements of what negotiators often refer to as creative ambiguity. This is not to say that outcomes of institutional bargaining are bound to prove antithetical to the pursuit of good governance, so that those responsible for implementing and operating regimes will invariably find themselves confronting more or less serious problems relating to institutional interplay at the outset. Rather, forces governing formative links are distinct from those that arise in connection with operational links, a fact that makes the relationship between these two types of institutional interplay somewhat haphazard.

Operational Links

Whereas formative links center on institutional interplay arising in the course of regime (re)formation, operational links involve processes characteristic of efforts to move regimes from paper to practice and especially to administer or operate these arrangements successfully on a day-to-day basis. Unlike contracts that are self-executing in the sense that they do not result in continuing relations among the parties, regimes must be tended on a regular basis to be effective. Depending on the defining characteristics of the arrangements in question, managers will find themselves dealing with matters relating to achieving compliance with rules, fulfilling commitments made by members, operating collective-choice procedures, developing programmatic initiatives, funding activities conducted under the auspices of the regime, and resolving disagreements regarding application of the regime's provisions to specific situations. Since these concerns require attention throughout the life of a regime, it is a matter of some importance to examine interactions between or among regimes regarding such matters; that is, to think about the politics of institutional linkages arising in operational settings.

Supplying Common Services

In many social settings it is helpful to draw a distinction between regimes themselves—understood as constellations of rights, rules, and relationships—and organizations—construed as material entities—established to provide various services necessary to operate these arrangements effectively every day.[6] A natural tendency in situations of this sort is to favor development of organizations that can provide services at one and the same time for a number—sometimes a large number—of distinct regimes. This leads immediately to the development of operational interactions among regimes that may have no significant functional links to one another but that are connected through ties to the same service organizations.

Consider the supply of financial resources, operation of compliance mechanisms, and provision of dispute settlement services. Many environmental and resource regimes operating in domestic settings have no independent source of funding; they rely on resources allocated from a general fund to carry out their operations. Conversely, revenues generated by the operation of such regimes, in such forms as grazing fees, royalty payments accruing from extraction of minerals or hydrocarbons, severance taxes, and pollution charges, flow into the general fund in many systems. The same goes for compliance mechanisms. Many regimes leave monitoring of conformance with regulatory requirements and collecting evidence pertaining to alleged infractions to agencies designated specifically to provide these services. As a result, they may have no more than rudimentary capabilities of their own in these areas. Much the same can be said about dispute settlement services. Individual regimes often depend on specialized agencies, typically located in ministries of justice, both to prosecute cases against subjects alleged to have violated their rules and to organize a defense when subjects claim that a regime has dealt with them unfairly or in a manner that violates the intent of its rules and regulations.

It is easy to see that this separation of regimes and organizations that arise to service them often leads to linkages between and among distinct regimes. Sometimes this is simply a matter of competition for scarce resources or finite capacities. Funds allocated to operate one regime may come at the expense of resources to operate other regimes. Compliance mechanisms preoccupied with issues relating to violations of the rules of one regime may be unable to pay adequate attention to questionable activities occurring in connection with other regimes. On the other hand, in some cases linkages of this sort undoubtedly enhance the effectiveness of all or most of the regimes involved. To begin with, organizations specializing in matters of funding, compliance, or settling disputes can improve efficiency by taking advantage of economies of scale. They also can profit from lessons learned in specific areas that can be generalized to other areas and by taking advantage of opportunities to experiment on a limited basis with innovative procedures that can be applied in a number of areas once they are approved for general use. These observations do not yield a general formula regarding the optimal scope and scale of these operational links, but they help account for the ubiquity of such linkages in many settings.

Some will react to this account by thinking that operational links are not likely to be of much importance at the international level, however widespread and significant they may be in domestic settings. The reasoning is straightforward. International society lacks a government or public authority of the sort required to establish and administer service organizations dealing with matters such as funding, compliance, and provision of dispute-settlement services on a centralized basis. It follows that international regimes created to solve specific problems typically have to provide their own services, a situation that both sets them apart from their domestic counterparts and imposes a significant burden on those responsible for their administration. Yet it would be a mistake to carry this line of reasoning too far. Domestic regimes sometimes prove resistant to the efforts of organizations to provide services on a centralized basis, and a number of interesting experiments are now under way that point to prospects for growth of operational links at the international level during the foreseeable future.

Legislation creating regimes at the domestic level ordinarily designates a lead agency for each arrangement and invests these agencies with the authority to administer individual arrangements day to day. In the United

States, for instance, the Forest Service is responsible for administering rules governing human activities in national forests, the National Marine Fisheries Service administers the regime dealing with harvesting fish in the exclusive economic zone, and the Bureau of Land Management handles arrangements governing grazing on lands that are part of the public domain. None of these agencies is in a position to become a wholly autonomous body, generating its own funds or dealing with violators in whatever way it sees fit. Yet lead agencies often become highly protective of their own turf. It is not uncommon for them to set up their own monitoring systems, establish their own compliance mechanisms, and, in general, resist interference on the part of outsiders. In some cases, lead agencies even seek to develop revenue sources of their own, although legislative committees and executive bodies (e.g., the U.S. Office of Management and Budget) can be counted on for obvious reasons to try to suppress such moves. None of this diminishes the importance of organizations designed to provide joint services at the domestic level. But these observations do indicate a need for caution in contrasting domestic and international settings in these terms.

Equally important is the proliferation of experiments involving joint services at the international level. Some of the resultant arrangements are now well established. The U.N.'s specialized agencies such as the IMO supply administrative services for a number of distinct regimes. In other cases, innovative arrangements are emerging under which individual regimes (e.g., ozone regime, the regime governing trade in endangered species, the regime dealing with transboundary movements of hazardous wastes) have their own secretariats that are loosely connected under the umbrella of a single organization (e.g., UNEP). There is a considerable history of efforts to create separate organizations to supply scientific assessments required by those responsible for the administration of specific regimes. One of the best-known examples is the International Council for the Exploration of the Sea, which was founded at the beginning of the twentieth century and now plays an acknowledged role in supplying credible scientific assessments to a number of international fisheries regimes. A more complex and ambiguous case involves the relationship between the Subsidiary Body on Scientific and Technological Advice, which is an integral component of the climate regime, and the IPCC, which is a separate body operating under the auspices of UNEP and the World Meteorological Organization.

Perhaps the most interesting recent development relating to the supply of joint services at the international level centers on the activities of the Global Environment Facility (GEF). This organization, operating under the joint auspices of the World Bank, UNEP, and U.N. Development Programme, has emerged as an important mechanism for funding programmatic activities launched in connection with the regimes dealing with climate change and biological diversity; it also provides funding to supplement the efforts of the Montreal Protocol Multilateral Fund to facilitate phasing out of ozone-depleting substances (Fairman 1996).[7] The GEF has proved controversial for several reasons, yet it cannot be denied that solidification of its role in recent years is a major step forward in the supply of joint services at the international level (Sand 1999).

There is no comparison between the scope of institutional interplay arising from the role of organizations supplying joint services to a number of distinct regimes at the domestic level and the parallel phenomenon at the level of international society. Even so, operational links of this sort are on the rise at the international level. When regimes dealing with matters as diverse as climate change, biological diversity, and ozone depletion are linked to each other because they share a funding mechanism, the scope for institutional interplay expands rapidly. Under the circumstances, it is reasonable to anticipate that the politics of institutional linkages will become increasingly prominent in this realm during the foreseeable future. Whether the results will, on balance, enhance the performance of individual regimes remains to be seen. But as controversies surrounding the role of the GEF make clear, political processes arising from operational links are a fact of life at the international level.

Reconciling Institutional Overlaps

Institutional overlaps occur regularly in all social settings; they can be expected to become more common and more significant as the density of institutional arrangements operating in the same social space increases. This explains the growing interest in overlaps among international regimes that deal with environmental problems on the one hand and with commercial or economic concerns on the other (von Moltke 1997). What makes overlaps particularly important and worthy of separate consideration in this discussion of operational links is the fact that they not only pose technical challenges to those seeking to avoid or minimize mutual interference in the day-to-day operations of individual regimes, but also generate more or less severe conflicts of interest among influential actors in the affected areas. Arguments about the compatibility of rules governing international trade and various environmental measures (e.g., restrictions on transboundary movements of hazardous wastes under the terms of the Basel Convention, restrictions on the development of genetically modified organisms under the terms of the biosafety protocol to the Convention on Biological Diversity) are not purely technical. They involve the interests of actors that are powerful in economic terms as well as environmental groups concerned with protecting public health and avoiding what has become known as eco-imperialism.

It follows that the development of effective procedures to resolve, or at least to manage, conflicts arising from institutional overlaps is a critical concern in every social setting. A number of approaches to this problem are in common use. In the private sector, where the issue is framed primarily in terms of efficiency or profit maximization in contrast to fairness or equity, overlaps are often reconciled through mergers brought about either by amicable agreements or by hostile takeovers. Where significant economies of scale are exploited through expansion of production, horizontal integration is an attractive option. Opportunities for reducing production costs either by increasing the compatibility of intermediate products or by ensuring that such products are supplied in a timely manner make vertical integration an attractive device. Whatever the virtues of these responses to overlaps in the private sector, however, they have little to tell us about the treatment of institutional overlaps when the problem arises from conflicts about rules of separate regimes in contrast to inefficiencies arising from efforts to reap benefits under the terms of regimes or constellations of rules that are not contested.

At the domestic level, courts and, in the final analysis, legislatures are primary mechanisms for resolving conflicts arising from institutional overlaps. Efforts to resolve conflicts between environmental regulations and rules governing economic activities have become a significant component of the case load of courts in developed countries such as the United

States. Do the rules articulated in the regime for protection of endangered species require suspension of contracts governing the harvest of timber in certain national forests or reconsideration of approved projects involving construction of dams on certain rivers? Do regulations promulgated to implement the regime for clean air require burning low-sulfur coal, regardless of the terms of preexisting contracts (Munton 1998)? How much discretionary authority do administrators have in applying the provisions of these regimes to specific circumstances that arise? As these examples suggest, those who are dissatisfied with the judgments that courts render often appeal to legislatures to modify the rules of existing regimes in such a way as to overturn or at least to revise those judgments. When regimes require periodic legislative reauthorization, such as arrangements established under the terms of the Endangered Species Act, the Clean Air Act, or the Fishery Conservation and Management Act in the United States, recurrent legislative battles are standard. In the wake of particularly controversial court decisions, these battles become heated; they often lead to adjustments in prevailing regimes that reflect the political influence of groups whose interests are affected.

Procedures for coming to terms with institutional overlaps are not guaranteed to produce outcomes that contribute to achievement of social goals such as sustainability, efficiency, or equity. But they are far more developed than their counterparts operative at the international level (Young 1999c). Still, the increasing frequency of overlaps in international society has led to marked growth of interest in this problem and to a variety of experiments with procedures designed to solve it. The most striking examples involve overlaps between regimes dealing with specific environmental problems and the more comprehensive arrangement governing international trade. The central debate concerns the (in)adequacy of the Committee on Trade and the Environment of the WTO and the (un)acceptability of the WTO's dispute-resolution procedures as devices for handling issues involving significant overlaps between environmental regimes and components of the trade regime. But other cases are worthy of consideration as well. It will be interesting to see, for instance, whether the recently established International Tribunal for the Law of the Sea can play a constructive role not only in reconciling overlaps among components of the clustered regime dealing with marine matters, but also in sorting out overlaps between arrangements included within the law of the sea and various environmental regimes (e.g., for biological diversity).[8] Similarly, it is possible to imagine significant roles for organizations such as the GEF in finding ways to deal with overlaps among different environmental regimes that focus on specific problems.

Despite these developments, ad hoc negotiation is likely to remain the predominant mechanism for handling overlaps in international society for some time. Recognizing this fact, individual interest groups will take steps to structure such negotiations in ways that favor their own causes (a phenomenon discussed in some detail in the next section with particular reference to overlaps between environmental and trade arrangements). The overall effect is to reduce the role of specialized organizations (e.g., courts) in determining the course of the politics of institutional linkages at the international level. Some will surely see this as a defect in international society that has to be corrected sooner rather than later, especially in light of the continuing rise in density of international regimes. The magnitude of transaction costs arising from efforts to reconcile institutional overlaps through ad hoc negotiations is a strong argument in favor of this proposition. Yet there is no compelling reason to conclude that it makes sense to emulate the experience of domestic societies in the search for improved procedures. The performance of domestic systems measured in terms of criteria such as sustainability, efficiency, and equity is by no means reassuring. Much can be said, under the circumstances, for making a concerted effort to encourage innovation in the search for mechanisms to handle institutional overlaps that are adapted to particular prevailing conditions.

Strategic Uses of Interplay

Both formative links and operational links give rise to political processes. They cannot be understood fully in terms of conventional approaches to public administration or handled effectively through the application of simple procedures based on some ideal conception of good governance or the pursuit of the public interest. Nonetheless, these links do feature efforts motivated in considerable part by a desire to create arrangements that will prove successful in solving the problems that led to their formation, and to administer these arrangements in such a way as to ensure success. Exploitation of institutional interplay for strategic purposes, in contrast, occurs when actors strive deliberately and predominantly to take advantage of institutional overlaps to pursue their own agendas. For those engaged in such strategic actions, the extent to which regimes solve problems whose importance is acknowledged by all parties becomes a secondary consideration. Instead, the center of attention shifts to efforts on the part of major actors to exploit interactive decision making to promote their own ends regardless of the consequences in terms of the common problem.

Engaging in Institutional Foreplay

Those who approach linkages strategically sometimes engage in institutional foreplay in the sense of seeking deliberately to adjust or alter existing institutional arrangements in order to reduce asymmetries and strengthen their hand in subsequent efforts to (re)form institutions. Perhaps the most obvious cases arise under conditions featuring interactions between a single regime that is perceived as large and powerful and a collection of more narrowly defined regimes that are thought to be weaker. But other forms of institutional foreplay are certainly imaginable. Thus, it may make sense to take steps to strengthen a regime's compliance mechanisms or to improve procedures for settling disputes about the requirements of a regime's rules in anticipation of interactions with other regimes dealing with functionally related problems. The rationale for taking such steps would emphasize the importance of putting one's own house in order in preparation for interactions with others. Thus, the need to deal with external pressures may generate positive side effects with regard to the effectiveness of individual regimes within their own domains.

The most prominent current example of institutional foreplay relating to large-scale environmental concerns centers on the debate over creating a World Environment Organization (WEO) treated as a counterpart to the existing WTO. Advocates of a WEO have articulated a number of reasons for taking this step (Biermann 2000). One of the most persuasive goes like this. Many specific environmental regimes, such as those pertaining to ozone depletion, transboundary movements of hazardous wastes, and trade in endangered species, have provisions that deal with trade and that raise questions about the compatibility between trade measures embedded in environmental regimes and provisions of the over-arching international regime governing trade. It follows that interactions between such regimes will occur with some frequency. As things now stand, however, environmental regimes find themselves at a distinct dis-advantage in these interactions. The trade regime is a comprehensive ar-rangement that includes provisions (e.g., Committee on Trade and the Environment) designed to deal with environmental issues in a manner acceptable to the trade community and that is administered by an organi-zation—the WTO itself—whose capacity and resources dwarf those available to individual environmental regimes (Esty 1993). On this ac-count, a primary function of a WEO would be to level the playing field in these interactions. As long as the capabilities of the environmental com-munity are fragmented among a number of individually weak arrange-ments, proponents of a WEO maintain, the WTO will be able to divide and conquer in a manner yielding results detrimental to the pursuit of environmental goals. The creation of a WEO would reduce these asym-metries and thus contribute to the achievement of outcomes more in line with environmental goals.

Is this position persuasive? In essence, it rests on two propositions: fragmentation is a source of weakness in interactions between environ-mental regimes and the trade regime, and creation of a WEO would strengthen substantially the environmental side in these interactions. The validity of both claims is less than self-evident. International practice is moving toward acceptance of multilateral initiatives involving trade mea-sures to come to grips with environmental problems. To take a prominent example, the Montreal Protocol grants the Conference of the Parties au-thority to impose trade sanctions on individual members that fail to comply with accepted phase-out schedules. If a number of states that are influential members of both the ozone regime and the trade regime choose to support these sanctions as a legitimate application of the provisions of the ozone regime, there is no reason to suppose that supporters of the trade regime will be able to overturn this initiative on the grounds that it conflicts with rules of the trade regime. Note also that creation of a WEO could prove costly in terms of both material resources required to run a sizable environmental organization and emergence of turf battles between those whose loyalties lie with individual environmental regimes and those who favor integrating individual environmental regimes under the auspices of a WEO. As a result, the effort to establish a WEO could prove divisive rather than strengthening the hand of environmental regimes in dealing with the WTO. Clearly, institutional foreplay is an issue to be reckoned with in connection with the larger subject of interplay between economic and environmental regimes at the international level. But the ultimate outcome of this strategic interaction is far from clear.

Devising Linkage Strategies

Turn now to consideration of options available to those who attempt to take advantage of existing regimes to promote their own agendas. Needless to say, exploitative measures must be tailored to the circumstances prevailing in specific situations if they are to yield a high probability of success. Yet it is helpful to differentiate several families or constellations of linkage strategies that are easy to identify and often used. Three worthy of consideration here are institutional capture, institutional reform or repeal, and institutional integration.

In some situations, strategic interaction centers on efforts on the part of actors that are initially outsiders to exploit linkages in such a way as to capture regimes expected to affect their interests significantly. Sometimes this is essentially a matter of finding ways to control the selection of arenas in which regimes are formed in the first place. The struggle between proponents and opponents of UNEP as an appropriate arena for creation of the climate regime, referred to earlier in this chapter, is a prominent case in point. But launching institutional initiatives designed to capture regimes already in place is an equally important phenomenon. Those who worked hard during the 1980s to persuade the International Whaling Commission (IWC) to pass a moratorium on harvesting great whales and who opposed implementation of the Revised Management Procedure (RMP) sought to shift the basic agenda of the regime from conservation to preservation. Similarly, one objective of those advocating creation of the Arctic Council was to annex the preexisting Arctic Environmental Protection Strategy and to balance the environmental protection program operating under its auspices with a newly created sustainable development program mandated in the 1996 Ottawa Declaration establishing the Arctic Council (Scrivener 1999).

Under other conditions, those who find existing regimes threatening may abandon efforts to alter the provisions of these regimes and try instead to replace them with more acceptable alternatives. The history of efforts to regulate whaling is an interesting example of this type of strategy. Frustrated by unwillingness of the IWC to lift the blanket moratorium on harvesting whales and to proceed with implementation of the RMP, whalers in the Faroe Islands, Greenland, Iceland, and Norway established the North Atlantic Marine Mammals Commission (NAMMCO) in 1992 as a mechanism for protecting and promoting their interests (Hoel 1993). To be sure, these actors have always maintained—at least in public—that NAMMCO should not be considered a potential alternative to the IWC; its objective is merely to improve management practices relating to marine mammals, including but not limited to whales, in the North Atlantic area. But the strategic significance of this initiative is clear. Iceland withdrew formally from the whaling regime. Norway states that it is not bound by the general moratorium with regard to harvesting minke whales. The Faroes and Greenland take the view that the strictures of the IWC should not apply to them. Members of NAMMCO are under no illusion about their ability to capture the whaling regime. But they are clearly signaling that they may go their separate way if the IWC does not alter its stance regarding harvesting of non-endangered whales (Caron 1995). The implicit threat is that those whose actions ought to be regulated under the terms of the RMP will take their business elsewhere if the IWC does not change its ways, thereby calling into question the legitimacy and even the relevance of the existing whaling regime (Friedheim 2001).

The third family of linkage strategies centers on efforts to rationalize existing institutional arrangements through various forms of integration. International institutions, like institutional arrangements in other social settings, have a tendency to evolve haphazardly in the absence of a master plan. This may result from initiatives involving overlapping but not identical geographical areas (e.g., Arctic Council and Barents Euro-Arctic Region), related but not identical functional concerns (e.g., environmental protection and sustainable development), or advent of overarching arrangements that may subsume preexisting regimes (e.g., the regime for biological diversity and the more specific regime covering trade in endangered species). In such cases, some major players may find it attractive to launch initiatives to rationalize and even integrate collections of institutional arrangements that have become increasingly difficult to disentangle in practice. Naturally, those who take the lead in this area are likely to claim that their actions are motivated by considerations of good governance and the need to come to terms with costly institutional overlaps. In specific cases, such claims may contain an element of truth. Yet it is clear such initiatives are commonly energized by a desire to promote the interests of their principal supporters. It may make sense from a managerial perspective, for instance, to look for ways to link fisheries regimes whose purpose is to regulate harvesting of targeted species and the overarching biodiversity regime that seeks to maintain the integrity of large marine ecosystems (Hoel 1999). But it is also clear that the focus on whole ecosystems and the associated tendency to embrace the precautionary principle, which are prominent features of the biodiversity regime (Sand 2000), will lead to management decisions that sometimes conflict with preferences of harvesters whose interests normally dominate the activities of fisheries regimes (Norse 1993). Of course, outcomes arising from integration strategies in particular instances will be affected by the politics of relevant issue areas. But it seems indisputable that interest in matters of this sort will grow as the density of institutional arrangements operating at the international level increases.

Future Directions

Institutional interplay becomes a matter of politics when actors engaged in specific interactions seek consciously to make use of overlaps to achieve identifiable goals. Sometimes this is a matter of designing and administering distinct institutions in such a way as to solve common problems, or to enhance social welfare or promote the common good. Situations of this kind give rise to a lively interest both in the initial development of institutional arrangements and in the creation of appropriate organizations to administer or manage the resultant arrangements on a continuing basis. In other cases the emphasis falls on exploiting institutional interplay in such a way as to promote the interests of particular actors, whether or not this contributes to problem solving. The politics of institutional linkages normally feature a complex mix of efforts to enhance social welfare and to promote the interests of individual participants. This is likely to confound the efforts of those who want to design efficient and equitable regimes to solve problems without paying attention to conflicts of interest that constitute a common feature of this form of interactive decision making. Yet mixed motives are familiar to analysts who study strategic interactions in a variety of settings and who have accumulated considerable intellectual capital that can be brought to bear in efforts to understand the politics of institutional linkages (Schelling 1978).

This chapter and its predecessor have explored issues relating to horizontal and vertical interplay on the assumption that they can be separated and analyzed independently. Surely, this makes sense as a research strategy. Nonetheless, the two forms of institutional interplay often occur simultaneously and interact with each other in ways that affect collective outcomes significantly. The creation of a WEO and the effects of such an initiative on interactions between environmental and trade regimes at the international level, for instance, would undoubtedly have repercussions affecting cross-scale interactions between international regimes and institutions that operate at the national level and that are likely to be assigned the task of implementing specific international agreements. Among other things, such a development would trigger a range of issues involving relations between ministries of trade or commerce and ministries of the environment at the national level, which can be shunted to the back burner as long as links between environmental and trade regimes are not regarded as priority concerns or recognized under terms of explicit international agreements. It follows that links between horizontal and vertical interplay are destined to become a topic of considerable interest to those who search for ways to come to terms with large-scale environmental changes.

6

Scale: Addressing Local and Global Environmental Challenges

Can we apply lessons derived from the burgeoning literature on the management of common-pool resources (CPRs) in small-scale, often local settings to the treatment of global challenges involving such matters as depletion of the stratospheric ozone layer, loss of biological diversity, and changes in Earth's climate system? Advantages arising from an affirmative answer to this question are apparent, and a number of those who have studied small-scale systems have reached optimistic conclusions about the prospects for transferring propositions from one level to another (McGinnis and Ostrom 1996; Burger et al. 2001). In a major article published in *Science,* five prominent contributors to the literature on long-enduring CPR institutions conclude, "Some experience from smaller systems transfers directly to global systems," although they are careful to note that "global commons introduce a range of new issues, due largely to extreme size and complexity" (Ostrom et al. 1999, 278).

How should we evaluate propositions of this sort? More specifically, how can we differentiate conclusions derived from study of small-scale systems that do apply at the global level from those that do not? In this chapter, I propose that this question centers on the problem of scale, an issue that has received comparatively little attention among social scientists, although it is a prominent concern throughout the natural sciences (Young 1994b). The key issue is easy to articulate. Are large-scale systems essentially macrocosms of small-scale systems, so that it is a straightforward matter to scale up findings derived from the study of the latter or, for that matter, to scale down findings based on the study of the former? Or are more or less significant differences present in the structures and processes of systems operating at different scales, so that assuming transferability of propositions across spatial or temporal scales—in this case, levels of social organization—is a risky business? The attractions of assuming that it is possible to transfer propositions are evident. But as many disappointed scientists, not to mention policy makers who have followed their lead, have discovered, the pitfalls associated with scale awaiting the unwary are numerous and often debilitating.

This chapter focuses on matters of spatial scale with particular reference to the relationship between management of small-scale CPRs and handling of global environmental challenges. In considering the prospects for scaling up, I focus on problem structure, or the nature of the problems arising in the two settings; agency, or the character of actors in small-scale systems and in international society; and social context, or the nature of social settings prevailing at the two levels. A concluding section comments briefly on design implications of differences between small-scale CPRs and global environmental problems. Given the nature of the issues, I move back and forth between local concerns (e.g., sustaining specific fish stocks) and global challenges (e.g., climate change) to illustrate the arguments presented.

Nothing in this analysis suggests that we should eschew efforts to scale up from small to large systems or, for that matter, to scale down from macro to microsystems. Although the literatures on local, community-based resource regimes and on international environmental regimes are both expanding vigorously, contact between them is surprisingly scarce. Under the circumstances, much is to be gained and little is to be lost from a sustained effort to compare and contrast the findings of the two streams of analysis. Even so, the principal conclusion of this chapter is that differences between the two sets of cases are substantial. The pitfalls confronting those desiring to scale up in this context are correspondingly severe; incautious efforts to apply local lessons to global challenges may do more harm than good.

Problem Structure

Common-pool resources are conventionally defined in terms of two dimensions: excludability and rivalness (or subtractability). If we treat each dimension as a dichotomy, it is possible to construct a 2 × 2 table that makes it easy in conceptual terms to identify CPRs and to differentiate problems associated with them from problems with other structures. The CPRs are goods that are nonexcludable in the sense that it is costly, or even impossible, to prevent all members of a group from enjoying their benefits once they are available to any member, and rival (or subtractable) in the sense that consumption of a unit of these goods on the part of any individual member diminishes or degrades the supply available for consumption or appropriation on the part of others (table 6.1).[1] The CPRs differ from pure public goods, which are nonrival as well as nonexcludable, from private goods, which are excludable and rival, and from club goods, which are excludable and nonrival (at least among members of the relevant club).[2]

The fundamental challenge arising in connection with CPRs is to avoid or overcome the familiar tragedy of the commons, in which a group of appropriators, acting on the basis of individualistic calculations, deplete resources or use them unsustainably due to lack of incentives to husband resources in a setting where the benefits of conservation will accrue largely to others. Human uses of a variety of living and flow resources, such as fish, forests, pastures, and aquifers, often pose such problems. Efforts to solve these problems are the main focus of the rapidly growing literature on long-enduring CPR institutions in small-scale societies (Ostrom 1990). With a little ingenuity, it is possible to expand the scope of this analysis by treating ecosystem services (e.g., capacity of watersheds and airsheds to absorb human wastes) in somewhat similar terms. It is often hard to exclude individual users of these services from dumping wastes into water bodies, for instance, even though at some stage these practices will degrade or even overwhelm the capacity of the ecosystems to handle wastes.

Table 6.1
Types of goods and resources

	Excludable	Nonexcludable
Rival	Private goods	CPRs
Nonrival	Club goods	Pure public goods

CPRs = common-pool resources

To what extent do international and ultimately global problems involve efforts to manage CPRs? It turns out that this seemingly simple question harbors significant complexities. Most important, environmental problems are socially constructed in the sense that there are almost always a number of plausible ways to think about them, and the choice of conceptualizations is likely to have significant consequences for the interests of one or more members of the relevant group (Jasanoff and Wynne 1998). Efforts to distinguish between excludable and nonexcludable resources are a case in point. In most situations, excludability is a human artifact rather than an unalterable natural condition. Fish stocks, for instance, are often seen as nonexcludable resources. But managers all over the world are experimenting with a variety of techniques, such as individual transferable quotas, to create effective exclusion mechanisms in this realm (Iudicello, Weber, and Wieland 1999). The same is true for water, with a long history of efforts to achieve excludability by developing various types of water rights. A particularly interesting case in today's world involves exclusion mechanisms relating to waste disposal through the creation of discharge permits covering liquid wastes and emissions permits applying to gaseous wastes. The purpose of these observations is not to advocate a particular type of exclusion mechanism or even to contend that creation of such mechanisms is necessarily a good thing. Rather, the point is that the extent to which any given environmental concern is properly construed as a CPR problem is likely to be more a matter of how we look at it than a fact of life.

Even so, it is possible to offer some relevant observations about the character of environmental problems arising at the international level and to compare and contrast them with small-scale ones. Some large-scale issues do lend themselves to treatment as CPR problems (Barkin and Shambaugh 1999). Examples are consumptive uses of birds and marine mammals (e.g., whales) that migrate across jurisdictional boundaries, and fish stocks that transcend the boundaries of national jurisdictions (e.g., highly migratory species such as tuna). It is also true with regard to appropriation of certain nonliving resources whose supplies are finite, such as the electromagnetic spectrum and deposits of manganese nodules located on the deep seabed.[3] In such cases, it seems reasonable to conclude that the fundamental character of the problems is essentially the same whether they are local or global. It is no surprise, then, that many efforts to create and implement international resource regimes bear a distinct resemblance to similar efforts in small-scale settings.

In other cases, however, the relationship between classic CPR problems and global challenges is less straightforward. Many global challenges that occupy our attention are not consequences of the actions of a group of relatively homogeneous appropriators seeking to maximize individual benefits accruing from use of the same resource. Rather, they take the form of externalities or unintended side effects of efforts to pursue other objectives. No one sets out intentionally to destroy stratospheric ozone or to alter Earth's climate system; no one benefits by beating others to the punch and appropriating the lion's share of these resources. True, the stratospheric ozone layer and the climate system can be thought of as ecosystems producing services that are nonexcludable and depletable. But in many cases, those whose actions produce the relevant externalities are able to shield themselves from the negative results of their actions either by removing themselves from the impact zone (e.g., locating their homes upwind from smoke stacks) or by insulating themselves from relevant effects (e.g., installing air conditioners with effective filters). In such cases, the task of regulators is not a matter of persuading appropriators to join forces in devising common restraints from which they themselves will benefit directly. Rather, it is a matter of society at large imposing restraints on polluters who are likely to view these restraints as costly limitations that, at best, impose similar limitations on their competitors, and, at worst, disadvantage them in competition focused on the production of other goods and services. Where substantial asymmetries exist among competitors, as in the case of operators of coal-fired versus hydroelectric power plants, treatment of these regulatory issues is likely to become especially controversial. Operators of coal-fired power plants, who are subjected to costly regulations designed to reduce sulfur or carbon dioxide emissions, are likely to object that this puts them at an unfair disadvantage in relation to operators of hydroelectric power plants, who are not required to shoulder the costs of repairing damages that dams inflict on ecosystems.

Another set of cases encompasses situations that turn on conflicts among users possessing divergent lifestyles and seeking to exploit the same resource for different purposes. In extreme situations, these problems degenerate into intractable value conflicts, as in the confrontation between conservationists who wish to harvest whales in a sustainable manner and preservationists who regard any consumptive use or killing of whales as unacceptable and even immoral. In less extreme forms, conflicts among groups seeking to use the same resource for different purposes are both common and highly relevant to our global challenges. They occur among subsistence users of forest products, timber companies, owners of oil palm plantations, and those who want to preserve forests to protect biological diversity; between upstream users desiring to divert water for irrigation and downstream users interested in navigation, and between developers seeking to erect buildings in coastal areas and conservationists dedicated to protecting coastal wetlands and mangrove forests. In all these cases, the problems are more complex than those involved in persuading a group of more or less homogeneous appropriators to impose restrictions on themselves in the interest of conserving a resource that they all think about and value in similar terms. This is not to say that societies have no capacity to cope with situations of this kind. But to the extent that the solutions require value judgments or adjudication of claims based on fundamentally different classes of rights to the same resources, the problems are likely to prove more difficult to resolve than those envisioned in the ordinary CPR setting.

Going a step farther, we encounter problems that are essentially jurisdictional in the sense that they turn on ambiguities or disagreements about who has or should have authority to make decisions or promulgate rules regarding human uses of valuable resources or ecosystems. The difficulty of devising a management regime for Antarctica where some countries (claimant states) asserted jurisdictional claims and others (nonclaimant states) explicitly refused to recognize such assertions is a case in point. But other equally important problems reveal the same defining feature. Consider whether flag states or some entity representing international society (e.g., an International Seabed Authority as envisioned in the 1982 U.N. Convention on the Law of the Sea), should exercise authority over activities relating to deep-seabed mining, or whether states in which companies are incorporated or some international entity (e.g., the World Trade Organization) should be empowered to make rules about the production and consumption of chlorofluorocarbons (CFCs) or fossil fuels. In simple situations involving the use of CPRs, these issues do not arise. Members of the relevant group are appropriators, and their duty is to come to some workable agreement among themselves that applies to their own actions. With regard to many global challenges, however, human actions disturb ecosystems located wholly or partly beyond the jurisdiction of individual members of international society, and disagreements frequently arise as to the locus of authority to address resultant problems.

Three additional observations will help to flesh out this picture as it relates to environmental issues arising at various levels of social organization. First and undoubtedly foremost, some environmental problems, at the local as well as at the global level, are simply harder to solve than others. Some observers attribute this variation mainly to attributes of relevant biogeophysical systems and, in the particular case of global challenges, to issues of "extreme size and complexity" (Ostrom et al. 1999). Proceeding in this way, they contend that "[c]haracteristics of CPRs affect the problems of devising governance regimes" and point to a range of specific factors including the size and carrying capacity of the relevant systems, measurability of the resources, and the speed with which resources regenerate (Ostrom et al. 1999, 279). What should we make of this position in connection with prospects for scaling up from local concerns to global challenges? Some biogeophysical factors undoubtedly are important in dealing with global challenges (as well as with small-scale concerns). It makes a difference, for instance, whether populations of living resources are slow to regenerate (e.g., whale stocks) or are capable of bouncing back on the basis of one or two years of high recruitment (e.g., capelin or herring stocks). This is true also of the capacity of large ecosystems (e.g., the Great Lakes) to absorb human wastes without becoming severely degraded and the extent to which they have thresholds beyond which disturbances trigger irreversible changes.

At the same time, it is important to note that several attributes of human systems play key roles in determining how hard specific problems are to solve and that some of these concerns loom particularly large when it comes to meeting global challenges. To use language devised by students of international regimes, these factors are key determinants of the location of specific problems on a scale running from generally benign to highly malign situations (Miles et al. 2001). The number of human actors involved, the importance that key actors attach to the relevant activities, and the relationships of power among them are major considerations. Thus, the facts that DuPont produced about 25 percent of the world's CFCs in the mid-1980s, this product line still did not account for a major part of the company's overall business, and the company is more concerned with absolute gains than with relative gains, surely made the ozone problem easier to solve than the problem of climate change. Relevant activities regarding climate change involve large numbers of actors, go right to the heart of advanced industrial economies, and raise sensitive matters of North-South relations (Young 1999c, chapter 3). Similar remarks are in order regarding transparency, or the ease with which it is possible to determine what key players are actually doing in the relevant issue area. Many observers have expressed concern, for instance, that globalization will impede efforts to meet global challenges by making it increasingly difficult to monitor the behavior of multinational corporations with precision (Reinicke 1998). The strategic character of the problem or the structure of relationships among the major actors is another important consideration. As many analysts have pointed out, it is one thing to solve coordination problems (e.g., managing international air traffic) in which there are no strong incentives to cheat. It is something else to solve collaboration problems (e.g., managing consumptive uses of fish stocks) in which individual participants have incentives to cheat even though this is likely to lead to outcomes that are undesirable for everyone (Stein 1982).

A second observation concerns the role of uncertainty, a factor common to most environmental problems but often more severe in connection with large-scale systems. Although surprises are always possible, users of small-scale CPRs (e.g., in-shore and localized fish stocks) are likely to have a well-developed (although not necessarily scientific in the Western sense) understanding of the dynamics of biogeophysical systems in which they are embedded. Compare this with efforts to manage human activities in large marine ecosystems (LMEs), such as the Bering Sea or the South China Sea (Sherman 1992). The dynamics of LMEs typically feature chaotic behavior in such forms as ecological cascades or relatively sudden shifts from one state to another, which generates high levels of uncertainty as far as those responsible for operating resource regimes are concerned (National Research Council 1996). Among other things, this condition often allows self-interested members of user groups to find respected scientists ready and willing to make projections that suit their preferences, no matter what they are. This may be what Ostrom et al. (1999) had in mind in referring to the "extreme size and complexity" of many global problems. But note that the underlying issue is really a matter of uncertainty regardless of factors that cause it, and that the threshold between low and high levels of uncertainty will not always be a function of size

Finally, whatever the nature of a given environmental problem, development and implementation of a solution are likely to involve the supply of a public good. Thus, regulatory rules, decision-making procedures, and systems of implementation review tend to be both nonexcludable, in the sense that they apply to all members of the group, and nonrival, in the sense that applying them to one member of the group does not diminish or dilute their application to all members.[4] One way to think about this phenomenon is to say that creating a management regime to protect a natural resource or an ecosystem requires members of the relevant group to join forces to supply a public good, whereas the tragedy of the commons arises from actions of a group of users that deplete or destroy a good supplied by nature. The classic impediment to the supply of public goods centers on what is known as the free-rider problem (Olson 1965). Individual members of the group may derive net benefits from creating a management regime even if they have to pay their fair share of operating costs. But they will do even better in situations where they benefit from the operation of the regime while failing to comply with the rules or avoiding payment of fees. This is not to say that groups, at any level of social organization, will be unable to overcome this impediment. But it does explain preoccupation with procedures for overcoming the free-rider problem among both practitioners and analysts concerned with creating resource or environmental regimes.

Where does this account of the nature of environmental problems leave us regarding prospects for scaling up from local lessons to global challenges? Clearly, they do not vitiate all efforts to transfer insights drawn from the study of CPRs in small-scale settings to the assessment of global problems. But they do justify the conclusion that we should proceed with extreme caution. There is considerable room for debate and even conflicting conclusions regarding proper ways to think about most environmental problems, small-scale as well as large-scale ones. We have good reasons to doubt the wisdom of approaching every environmental concern exclusively or even primarily as a CPR problem. Although this conclusion applies at all levels of social organization, it is probably fair to say that we should be particularly careful to avoid assuming unreflectively that large-scale and especially global issues are properly treated as CPR problems.

Agency

If the nature of the problems underlying global challenges is one determinant of the extent to which we can scale up from small-scale to global systems, the character of the actors engaged in efforts to deal with these problems is another. Three questions must be addressed in this connection. First, assuming that those attempting to avoid or solve environmental and resource problems are unitary actors, what can we say about forces that drive their behavior at different levels of social organization? Second, what consequences flow from the fact that many actors involved in efforts to deal with global challenges are complex entities whose actions reflect internal dynamics of the sort envisioned in the concept of two-level games? Third, what are the implications of the fact that actors who become members of international regimes created to deal with global challenges often are not the same as those whose behavior gives rise to large-scale problems?

The literature on small-scale societies focuses almost exclusively on the behavior of individuals who are appropriators or users of resource(s), who must work together in the search for ways to avoid the tragedy of the commons, and who can be thought of without serious complications as unitary actors seeking to pursue their own goals under conditions of interactive decision making. Beyond this common core, however, it is possible to identify two separate streams of work on the management of CPRs that differ in the assumptions they make about the sources of the behavior in question. The essential difference between these streams is captured in the distinction that a number of analysts have made between the logic of consequences and the logic of appropriateness (March and Olson 1998). Thus, one school of thought, exemplified in most respects by the seminal work of Elinor Ostrom, conceptualizes issues relating to the management of CPRs as collective-action problems, treats members of groups of appropriators as self-interested utility maximizers, and approaches the issue of management as a matter of providing individual users with incentives to act in ways that conserve relevant resources (Ostrom 1990). The other school of thought, represented by anthropologists and sociologists such as Svein Jentoft and Bonnie McCay, directs attention to evolution of social practices that constrain actions of individual users, emphasizes the role of social norms, and envisions a significant role for socialization and other processes that lead participants to take norms and rules for granted in the absence of utilitarian calculations about the pros and cons of complying with these guidelines (Apostle et al. 1998; McCay and Acheson 1987). Some analysts, including Ostrom (1998), have begun to look for ways to combine or integrate these perspectives. For present purposes, however, it will help to consider them separately.

A sizable subset of those concerned with global problems build their analyses on the assumption that we can treat participants in international management systems or, to use a term that has become common currency in this field of study, international regimes as unitary actors (Levy, Young, and Zürn 1995). Within this group, it is undoubtedly correct to say that models based on utilitarian assumptions are the dominant approach. For those who adopt this premise, the problem of scale centers on similarities and differences between the way in which individual users in small-scale systems and members of international regimes in large-scale systems behave with regard to efforts to avoid or solve collective-action problems. Clearly, similarities between the two sets of analyses are substantial. Yet in a number of areas utility-maximizing members of international regimes—usually although not always thought of as states—may differ from their counterparts at lower levels of social organization. Some analysts hold that international actors pay more attention than their counterparts at the local level to the pursuit of relative gains in contrast to absolute gains (Baldwin 1993). Many believe that because states survive over the long run, they will employ lower discount rates (sometimes called social discount rates) than individuals in considering long-term consequences of current actions. Still others consider states to be less risk averse than individuals in dealing with specific environmental problems on the grounds that they are less likely to face ruin than individual appropriators as a consequence of making bad decisions in single-issue areas. Those working in this subfield have devised a number of additional propositions of this sort. But perhaps the key point to consider in the context of this analysis is that suggested differences can cut both ways with regard to their implications for avoiding or solving collective-action problems.

Although it remains a minority perspective, interest has grown in recent years in applying social-practice models to coping with global challenges. Here, questions that arise about differences between individual users of local CPRs and individual members of international regimes are more fundamental. What does it mean, for instance, to speak of norm-governed behavior on the part of collective entities such as states? Does it make sense to think in terms of socialization or development of a "habit of obedience," even when states are treated as unitary actors? What do we understand by the idea of social learning, and is learning on the part of states different from learning on the part of individuals (Social Learning Group 2001)? The point of posing these questions is not to argue that it is useless or even misguided to apply concepts such as socialization and learning to the behavior of states. After all, we are concerned with the construction of analytic models, which are normally judged in terms of their helpfulness or the correctness of their predictions rather than in terms of their descriptive accuracy. Rather, the question concerns the degree to which it is useful to transfer concepts such as socialization and learning, which were developed in efforts to understand the behavior of individuals, to studies dealing with the behavior of collective entities such as states. The jury is clearly out regarding this matter; studies of global issues that systematically draw on the intellectual capital of social-practice models are just beginning to yield results. Yet one likely outcome is that although these studies may generate useful insights, there is no one-to-one correspondence between individuals and large collective entities when it comes to behavioral consequences of the logic of appropriateness.

Beyond this, two larger issues relating to actors raise more profound questions about the prospects for scaling up from small-scale issues to global challenges. If we drop the simplifying assumption that states and other actors relevant to global challenges can be considered unitary, a whole range of issues come into focus that have no direct counterparts in studies of small-scale CPRs. To be sure, individual users of CPRs sometimes experience cognitive or even normative dissonance. Who has not felt pulled in several directions at once with regard to pros and cons of different strategies in situations involving interactive decision making? As a result, it would be naïve to suppose that individuals who initially agree to the rules of a management system based on some variation of common property will automatically comply with these rules. Much of the massive literature on compliance would be of no more than passing interest if this were the case.

Even so, one cannot compare the effects of cognitive and normative dissonance at the individual level, and the complications associated with internal dynamics of states and other actors engaged in efforts to meet global challenges. More specifically, at least three broad sets of issues surface once we relax the assumption that states are unitary actors. One set is captured in the idea of two-level games; it refers to the fact that states encompass numerous interest groups that may adopt conflicting views regarding the relative merits of international agreements, and often make concerted efforts to oppose implementation of specific agreements in domestic arenas (Putnam 1988). The inability of the Clinton administration to persuade the United States Senate to ratify the 1992 Convention on Biological Diversity and the 1997 Kyoto Protocol on climate change are dramatic illustrations of this. A second set of issues involves the organizational and bureaucratic processes involved in transferring provisions of international agreements from paper to practice. It is well known, for instance, that implementing rules often produces a substantial disparity between rules on paper and rules in use (Ostrom 1990; Victor, Raustiala, and Skolnikoff 1998). Yet another set of issues arises from the fact that governments, at least in democratic systems, must seek new mandates from their electorates from time to time. When elections bring new governments into power, the willingness of public agencies to live up to international commitments entered into by preceding governments may decline (or rise) sharply. Consider the attitude of the current Bush administration in the United States toward the Kyoto Protocol.

These last observations lead to consideration of what may well be the most fundamental difference between actors seeking to manage CPRs in small-scale settings and those engaged in efforts to meet global challenges. For the most part, actors who are appropriators of local CPRs and those who endeavor to regulate the actions of appropriators are the same. That is, the two groups are coextensive. It follows that avoiding or solving the tragedy of the commons at the local level is fundamentally a matter of self-regulation. Contrast this with the situation facing those concerned with global problems, where regulation is a two-step process. Those whose actions cause emissions of greenhouse gases, for example, include private corporations whose production processes are energy intensive, municipal authorities that own and operate power plants, and large numbers of individuals who drive cars and burn fossil fuels to heat their homes. Signatories to the 1992 U.N. Framework Convention on Climate Change and the 1997 Kyoto Protocol, in contrast, are states. Of course, in some instances rules established to cope with global challenges apply directly to the behavior of public agencies. This is especially true in societies where states own large quantities of natural resources or control a sizable proportion of the means of production. In the typical case, however, a clear separation exists between those who formulate and adopt the rules and those who are the subjects of the rules.

What are the implications of this feature of efforts to address global challenges? Even when governments are generally honest and effective, it is often difficult for public agencies to induce subjects to alter their behavior to comply with rules of international regimes. This is due to a combination of factors, including difficulties in monitoring subjects' behavior, the ability of many actors (e.g., multinational corporations) to escape jurisdiction of individual governments, and resources available to some subjects to fend off or co-opt regulators in both legal and political arenas. Under the circumstances, it will come as no surprise that many management systems or regimes that seem attractive on paper turn out to have little capacity to solve problems in practice. And when we drop or modify the assumptions that governments are generally honest and effective, problems arising from the gap between regulators and subjects grow by leaps and bounds. As those who have studied the implementation of domestic policies have shown repeatedly, what may look perfectly good in the capital often falls flat in the provinces (Pressman and Wildavsky 1973). This problem is at least as severe in the case of commitments entered into at the international level as it is with regard to purely domestic policies.

Where does this leave us with respect to scaling up from efforts to deal with CPRs in small-scale settings to initiatives aimed at meeting global challenges? The discussion here should not be taken to mean that there is no hope for efforts to scale up. Especially in studies that treat states as unitary actors and that approach issues as collective-action problems, the prospects for scaling up seem promising. Discussions that direct attention to the question of whether there is a social rate of discount, for instance, implicitly and perhaps reasonably assume that small-scale processes and global processes have much in common. But when we move to other analytic perspectives, the prospects for scaling up seem far more circumscribed. Note that this is not fundamentally a matter of extreme size and complexity. Rather, the problem arises from differences between individuals and states and from the separation between those who formulate the rules and those who are subject to them. What emerges in the case of global problems are top-down initiatives intended to guide the behavior of those responsible for the relevant problems with all the difficulties attendant on processes in which an imbalance is present between top-down and bottom-up efforts. Of course, states can do some things to draw subjects into processes of regime formation and to give them a stake in the success of the resultant arrangements. But even under the best of circumstances, the fact remains that this feature should make us wary of simple efforts to apply local lessons to the handling of large-scale challenges.

Social Context

The preceding sections focus on the nature of environmental problems and the character of actors whose behavior gives rise to them without reference to features of social contexts in which they are embedded. But environmental and resource problems do not arise and play out in a social vacuum. Even in small-scale settings where arrangements governing the use of CPRs (e.g., hunting whales or herding reindeer by indigenous peoples) constitute a defining feature of the social landscape, complex cultural practices emerge that deal with a variety of human interactions not limited to the primary resource complex. At the international level, environmental challenges, even important ones such as climate change, make up only a small segment of the overall agenda, which also encompasses issues of security, economics, and human rights. What this means is that efforts to address global challenges not only compete for attention and resources with other concerns but also, and most important, unfold in a larger social context whose character has significant consequences for the design of management systems or regimes and for the effectiveness of these arrangements. How different are social settings prevailing in the typical small-scale society and in international society? What are the implications of these differences in applying local lessons to global challenges?

The answers to these questions depend in part on the conceptual and theoretical lenses worn by those who address them (Young 2001a). Those whose view is centered on collective-action models and especially the familiar constructs of game theory are likely to take the view that all relevant contextual matters are, or can be, endogenized in utility functions or in the information embedded in games in normal form. Others who think in terms of social-practice models and emphasize the logic of appropriateness generally expect that the behavior of actors involved in specific environmental problems will be affected substantially by a range of cognitive, cultural, and socioeconomic forces that are not specific to individual environmental problems (March and Olsen 1998). Those who are optimistic about the prospects for scaling up typically view the world more through the lens of collective-action models than in terms of social-practice models. Of course, there are obvious analytic attractions to a strategy that seeks to come to terms with environmental problems without reference to matters belonging to an open-ended category such as social context. Yet this does not mean that considerations of context are irrelevant.

Those who want to avoid the tragedy of the commons at the local level often rely on several mechanisms of social control to strengthen resource regimes that are underdeveloped (some would say nonexistent) at the global level. Rules and decision-making procedures devised to regulate the actions of users of small-scale CPRs are normally embedded in broader cultural systems. Significantly, this link to culture has a cognitive as well as a normative component. Common understandings of forces at work in key ecosystems as well as probable results of human uses of relevant resources often become elements of the shared intellectual capital that guides the thinking of members of user groups. These common understandings may turn out to be incorrect in specific cases. But the point is that culturally embedded understandings are generally taken for granted in the sense that they prescribe normal or natural forms of behavior. Thus it is not necessary to worry about finding ways to induce individual participants to comply with prescriptions implicit in these understandings. There is simply no serious alternative to engaging in the familiar practices. What is more, when issues of compliance do arise, cultural practices commonly provide a variety of methods for sanctioning noncompliant behavior that can prove effective even though they do not involve the sorts of procedures associated with efforts to deter or punish violators in contemporary Western societies. Classic examples involve the operation of taboos, which can guide behavior effectively without regard to the logic on which they are based (Fienup-Riordan 1990).

It would be wrong to suppose that cultural encoding eliminates all problems arising from incentives to cheat or from efforts to maximize relative gains in small-scale settings.[5] Nor is it reasonable to conclude that no such thing as culture exists at the global level. The diffusion of cultural preferences, styles of thought, and even behavioral norms made possible by modern technology is a striking phenomenon. Nonetheless, it is difficult to identify a counterpart to cultural practices governing fishing, hunting, and herding in traditional small-scale systems operating at the global level. Is it possible to imagine the emergence of a global culture that would organize the ways in which we think about the protection of biological diversity or spell out norms relating to activities that generate emissions of greenhouse gases? Perhaps so. Yet to ask this question is to provide a clear sense of the magnitude of the gap between current realities and what would be required to develop cultural practices capable of being effective mechanisms of social control in efforts to meet global challenges. Although we should not dismiss this source of social control out of hand, it would be naïve to overlook the significance of the gap between small-scale societies and international society in these terms.

A somewhat related mechanism of social control that figures prominently in some analyses of small-scale CPRs centers on the role of community. It is not easy to devise a precise definition of community, even at the level of local societies. Yet most observers would agree that it has something to do with relationships between individuals and the larger groups to which they belong. Where community is strong, individuals derive a significant part of their identity from membership in the larger group, link their own welfare with some concept of social welfare, and are willing to subordinate the pursuit of narrow self-interest to achievement of the common good. Where community is weak, individuals are likely to care little for the welfare of the group and to resist pressure to subordinate enhancement of self-interest in the narrow sense to actions designed to promote the common good. It is easy to see how a strong sense of community can function as a mechanism of social control among users of CPRs (Singleton and Taylor 1992). To put the point in utilitarian terms, a sense of community gives rise to interdependent utilities, a condition that motivates individuals to consider the effects of their actions on the welfare of others rather than thinking exclusively in terms of their own welfare. Approached from a social-practice perspective, the same point can be thought of in terms of the incorporation of concern for the common good into the identities of individual members of the group. In either case, the result is the same. The tragedy of the commons is a product of highly individualistic or self-interested behavior occurring under conditions of interactive decision making. Community softens the impact of social dilemmas caused by behavior of this sort.

Can we rely on a similar sense of community to enhance the effectiveness of efforts to meet major global challenges? Many commentators refer, rather casually, to the existence of an international community. Yet it is difficult to see what this means in practice (Claude 1988). To the extent that we are concerned with international society in the sense of a society of states, the idea of an international community is difficult to interpret even in conceptual terms. If we are thinking about a global civil society, on the other hand, community is understandable in conceptual terms, but it seems clear that it is very weak at the international or transnational level (Wapner 1997). Does this mean that community has no place in efforts to meet global challenges? Perhaps the most interesting response acknowledges that community at the international level is presently weak but that signs indicate that it may be on the rise. Those who espouse this concept point to the apparent increase in the willingness of outsiders to incur costs in order to protect the human rights of those residing within the boundaries of individual states, as well as to the rise of ideas such as the common heritage of humankind to express a sense of community interest in protecting Earth's life support systems.

Even so, it is hard to avoid the conclusion that the nature of international society, at least in the familiar form of a society of states, is a significant barrier to the rise of a global culture or to the growth of a strong sense of community at the international level. Embedded in the concept of international society is a clear commitment to a set of norms and rules that grant states far-reaching authority over events occurring within their boundaries and that proscribe outside interference in their internal affairs except under extreme circumstances.[6] In effect, the extreme decentralization of international society and the vigorous opposition that proposals involving supranational arrangements typically evoke create a social setting in which the deck is stacked against those who want to bring broader mechanisms of social control to bear in efforts to solve specific problems.[7] This is why most efforts to create regimes dealing with global commons, such as the high seas or Earth's climate system, involve relatively elaborate rules intended to avoid infringements on the authority of member states (e.g., consensus rules regarding decision making or provisions allowing dissenters to exempt themselves from group decisions by entering objections or reservations).

At the same time, it is fair to say that we are witnessing significant changes in the character of international society, brought about at least in part by growing awareness that the costs of ignoring global challenges are likely to be great. Even in the realm of interstate relations, some initiatives featuring creation of international regimes (e.g., the ozone regime, the chemical weapons regime) have led to successful experiments with devices for overcoming or circumventing limitations arising from the general character of international society. Inspection procedures established to operate the chemical weapons regime, for instance, feature striking intrusions into domestic affairs of member states (Reinicke 1998). Beyond this, numerous signs (e.g., growing influence of a range of nonstate actors, emergence of a global civil society) suggest that developments are under way that are likely to produce substantial changes in the character of international society during the foreseeable future (Zürn 1999). We are reminded, under the circumstances, that the heyday of international society treated as a system of territorially based and sovereign states was relatively short. Does this mean that the familiar society of states with its elaborate protections for the rights of individual members is likely to collapse and be superseded by some alternative form of social organization at the global level? Not at all. What is probable is the emergence of a more complex social setting in which new actors and new social practices become relevant to the development of strategies designed to address global challenges. Whether this will, on balance, improve our track record in meeting global challenges is not easy to tell at this stage. At a minimum, however, these dynamic features of the social setting are sufficient to warn us against assuming that the context in which efforts to meet global challenges unfold is static or unchanging.

Stark differences can be seen between local and often traditional settings in which arrangements for managing the use of CPRs are hard-wired into cultural practices and the international setting construed as a society of states. But these are both ideal types that should not be allowed to rigidify our thinking, even though they are clearly helpful for some purposes. The most sophisticated literature on common-property systems recognizes that culture is not a magic mechanism of social control that can be counted on to guide the actions of users of CPRs to sustainable outcomes under all circumstances. The actual record regarding use of small-scale CPRs is replete with failures as well as successes (Krech 1999). Nor is the image of a society of states that are impervious to the effects of various mechanisms of social control an accurate reflection of the international setting today. Increasingly, states are allowing themselves to be drawn into more or less effective international regimes covering a variety of functionally and spatially defined problems. And some of the more interesting initiatives are not governed exclusively by the actions of states. Differences between small-scale settings and international society justify the proposition that we should proceed with extreme care in efforts to scale up. Yet it seems to me that we should not conclude from this that experiences dealing with CPRs and similar matters at the local level have nothing to contribute to our thinking about global challenges.

Design Implications

It seems logical to assume, as a first approximation, that considerable scope exists for scaling up and scaling down. Thus, the rapid growth of knowledge regarding the development and maintenance of sustainable management practices in small-scale systems seems to offer grounds for optimism. On second thought, however, differences between the two levels with respect to problem structure, agency, and social context appear significant. As a result, it is easy to come to the conclusion that local lessons have little to tell us in our efforts to meet global challenges. The message I wish to convey is that we should make a concerted effort to avoid both of these conclusions. It is unlikely that we will be able to scale up in any simple or mechanical sense. On the other hand, the observation that small-scale systems based on common-property arrangements can yield effective solutions to a variety of environmental problems is of great interest to those interested in parallel problems at the international level. In my judgment, adoption of this middle road offers the best hope for adding to our knowledge of environmental governance in human affairs.

Beyond this lies the question of the specific design implications of the argument set forth in this chapter. Of interest, many commentators (e.g., Ostrom et al. 1999) are not very precise in their comments regarding this matter. Ostrom and her colleagues observe that ". . . a scholarly consensus is emerging regarding the conditions most likely to stimulate successful self-organized processes for local and regional CPRs" (Ostrom et al. 1999, 281). But they content themselves with extremely general and somewhat imprecise formulations of these lessons, or, in their terminology, design principles, and they make almost no effort to discuss issues that arise in attempts to apply generic lessons to specific situations even at the small-scale level, let alone at the global level.

If the views set forth in this chapter are correct, they have a number of relatively specific implications for the design of resource or environmental regimes that are applicable not only to the way we respond to global challenges but also to how we treat environmental problems at other levels. Above all, we have to avoid the expectation that one size will fit all, or that we can devise a simple and unified checklist of steps that will allow us to deal with a wide range of problems in a more or less routine fashion (McGinnis and Ostrom 1996). Rather, we must consider matters of problem structure, agency, and social context and design management regimes on a case-by-case basis that are well suited to the circumstances. I can only scratch the surface of this topic here (see chapter 7 for a more general account of institutional design). Yet table 6.2 lists the principal distinctions regarding problem structure, agency, and social context and points to a number of design implications that arise.

With regard to problem structure, we can approach individual problems as CPRs, externalities, value conflicts, or jurisdictional differences. Although no characterization may be objectively correct, the choice of one perspective or another has substantial consequences. Whereas solutions to CPR problems call for self-regulation and efforts to build trust among appropriators, dealing with externalities often requires establishment of external regulatory mechanisms capable of applying formal sanctions to achieve high levels of compliance. Treatment of problems

Table 6.2
Issue attributes and design implications

Issue attribute	Design implication
Problem structure	
CPRs	Self-regulation, trust
Externalities	External regulation, sanctions
Value conflicts	Role of equity
Jurisdictional differences	Higher authority
Agency	
Unitary, complex actors	Targets of influence
(Non)utilitarianism	Role of incentives
Coextensive groups	One- and two-step processes
Social context	
Overarching culture	Socialization
Community	Appeals to the common good
Civil society	Social pressure

featuring value conflicts calls for responses rooted firmly and unambiguously in considerations of equity. Coming to grips with jurisdictional differences is likely to call for creation of a higher authority capable of making decisions about whole ecosystems, whether or not this takes the form of a formal organization of the sort we would normally regard as a government.

In the case of agency, we want to know whether actors are unitary or complex, the extent to which they base their decisions on utilitarian calculations of expected benefits and costs, and whether groups of users-appropriators and regulators are coextensive. It may be difficult to accomplish in practice, but it is a relatively straightforward matter to seek to influence the behavior of unitary utility maximizers through arrangements that affect their incentives. In cases featuring complex actors that respond to nonutilitarian cues or signals, efforts to solve environmental problems must address issues such as the development of appropriate discourses, social learning, and cultivation of feelings of legitimacy. And when groups of users and regulators are not coextensive, there is no substitute for efforts to cultivate feelings of legitimacy toward regulatory processes on the part of members of user groups.

With respect to the design implications of social context, where a strong, overarching culture is operative, it makes sense to design regimes that are compatible with and even build on culturally familiar mechanisms of social control and, in effect, to socialize regime members in terms of larger cognitive constructs and norms. The presence of a strong sense of community among members of the relevant group makes it possible to appeal to the importance of promoting the common good as a means of influencing the behavior of key actors. In the absence of community, there is no alternative to guiding behavior through measures that address the self-interest of individual actors. The existence of a civil society opens up prospects for using social pressure to influence the behavior of those whose actions give rise to environmental problems. At the international level, we are witnessing a steady rise in the role of various nongovernmental organizations in bringing pressure to bear on governments regarding environmental matters (Wapner 1996). Similar phenomena are discernible in efforts to come to terms with many problems in small-scale settings.

Are global challenges more difficult to meet than local challenges? Can we scale up lessons learned from experiences with small-scale CPRs in our efforts to address global challenges? In the end, I contend, these questions do not yield very insightful answers. Variance among environmental problems at all levels of social organization is sufficient to cast serious doubt on any effort to apply a single set of design principles without intimate knowledge of the character of specific problems and a concerted effort to construct arrangements well adapted to individual cases. Nonetheless, although small-scale systems and large-scale or global systems are far from identical, there is reason to believe not only that those seeking to solve global problems can benefit from the experiences of those dealing with local problems but also that those concerned with local problems stand to benefit from the experiences of those wrestling with large-scale problems. It follows that nurturing a dialogue between students of small-scale common-property regimes and students of international environmental regimes should be a matter of high priority for all concerned.

III

Analysis and Praxis

7

Usable Knowledge: Design Principles and Institutional Diagnostics

The pursuit of knowledge about the institutional dimensions of environmental change is worth while as an end in itself. Yet much interest in this subject arises from more applied concerns with producing usable knowledge, knowledge that will enhance our ability to control or regulate human actions that cause or threaten to cause disruptive, possibly irreversible, changes in human-dominated ecosystems (Vitousek et al. 1997). These range from small-scale systems that support limited groups of local appropriators to global systems that support all life on the planet (National Research Council 1999c). We want to know how and why some local appropriators have found ways to use renewable resources in a manner that is sustainable over time while others have failed conspicuously to do so, and whether we can apply lessons from these experiences to other cases (Ostrom et al. 1999). Similarly, we want to know why some international regimes are more successful than others in solving environmental problems and what we can learn from them about designing effective regimes to deal with a range of other problems (Young 1999a). In each case the underlying goal is the same, to derive lessons from past and present experience that can help us design institutions to deal with current and future environmental changes. Accordingly, the relationship between analysis and praxis is particularly important.

Under the circumstances, the sizable gap between the world of analysis and the world of praxis is cause for concern. Despite the efforts of a few individuals who have acquired the skills needed to operate comfortably in both worlds, most attempts to bridge this gap have yielded meager results. Researchers often frame their conclusions in abstract terms and publish their results in journals that practitioners seldom read. They emphasize uncertainties and are reluctant to make recommendations based on results of their work for handling specific environmental problems. For their part, policy makers commonly assume that it is difficult or impossible to come to terms with specific problems in the absence of detailed and continuously updated knowledge about the state of individual issues. They tend to look upon researchers as armchair analysts who have little contact with the actual course of events on the ground.

What can we do to bridge this gap and, in the process, improve the quality of the dialogue between these communities? In this concluding chapter, I address this question first by examining the effort to develop design principles, which is the most influential approach to the task of bringing analysis to bear on praxis in this realm. Then I discuss institutional diagnostics, which I portray as a distinct but by no means mutually exclusive approach. I propose that although the idea underlying design principles is highly attractive, there are serious limitations to what we can hope to achieve in the search for design principles applicable to a wide range of environmental concerns. The institutional diagnostics approach is less ambitious and therefore less attractive in ideal terms. Even so, I conclude that supplementing design principles with institutional diagnostics is likely not only to produce results that are more usable in dealing with specific problems but also to pave the way for a productive and continuing dialogue between researchers and policy makers in the search for institutional arrangements capable of limiting, if not eliminating, costly anthropogenic changes in human-dominated ecosystems.

Design Principles

The goal of developing generalizations that capture major findings about the institutional dimensions of environmental change and framing them in such a way that they can be applied to the handling of specific problems is relevant to all social settings. So far, however, those considering small-scale common-pool resources (CPRs) have made the most sustained and productive efforts to pursue this type of reasoning. As Elinor Ostrom, the leading contributor to this stream of analysis, puts it, a design principle is "... an essential element or condition that helps to account for the success of ... institutions in sustaining the CPRs and gaining the compliance of generation after generation of appropriators to the rules in use" (Ostrom 1990, 90). Basing their conclusions on a systematic examination of a sizable collection of actual cases, Ostrom and her colleagues have produced a set of specific design principles illustrated by long-enduring CPR institutions (table 7.1) (Ostrom 1990; Gibson, McKean, and Ostrom 2000).

In her account of these principles, Ostrom is suitably modest about the status of this endeavor, observing that she is ". . . not yet willing to argue that these design principles are necessary conditions for achieving institutional robustness in CPR settings" and that "[f]urther theoretical and empirical work is needed before a strong assertion of necessity can be

Table 7.1
Design principles illustrated by long-enduring CPR institutions

---

1. Clearly defined boundaries
   Individuals or households who have rights to withdraw resource units from the CPR must be clearly defined, as must the boundaries of the CPR itself.

2. Congruence between appropriation and provision rules and local conditions
   Appropriation rules restricting time, place, technology, and/or quantity of resource units are related to local conditions and to provision rules requiring labor, material, and/or money.

3. Collective-choice arrangements
   Most individuals affected by the operational rules can participate in modifying those rules.

4. Monitoring
   Monitors, who actively audit CPR conditions and appropriator behavior, are accountable to the appropriators or are the appropriators.

5. Graduated sanctions
   Appropriators who violate operational rules are likely to be assessed graduated sanctions by other appropriators, by officials accountable to these appropriators, or by both.

6. Conflict-resolution mechanisms
   Appropriators and their officials have rapid access to low-cost local arenas to resolve conflicts among appropriators or between appropriators and officials.

7. Minimal recognition of rights to organize
   The rights of appropriators to devise their own institutions are not challenged by external governmental authorities.

Source: Ostrom 1990, 90

made" (Ostrom 1990, 90). Nonetheless, the direction in which this line of reasoning is moving is clear. Ostrom states plainly that she is ". . . willing to speculate . . . that after further scholarly work is completed, it will be possible to identify a set of necessary design principles" (Ostrom 1990, 90–91). To the extent that it succeeds, the implications of this effort for policy making will be both easy to grasp and far-reaching. All those seeking to (re)form management regimes dealing with CPRs will know that the arrangements they create cannot endure or produce sustainable outcomes unless they include provisions that address concerns identified in each of the design principles.

Distinctive Features of Design Principles

What can we say about the character and content of design principles? For starters, it is apparent that Ostrom's principles emanate from an analysis inspired mainly by collective-action models in contrast to social-practice models. The analysis begins with consideration of familiar problems such as the tragedy of the commons, prisoner's dilemma, and the logic of collective action and proceeds to analysis of "the evolution of institutions for collective action."[1] Nothing in the generic concept of design principles requires a commitment to collective-action models. It is perfectly possible, for instance, that others could produce a set of principles highlighting sociological conditions involving such matters as feelings of legitimacy, the role of consensual knowledge, or the sense of community required to sustain various forms of diffuse reciprocity. Yet Ostrom's emphasis on conditions relating to collective-choice procedures, monitoring, sanctions, and conflict-resolution mechanisms is perfectly understandable in terms of thinking that rests on the assumptions characteristic of collective-action models.

Still, an element of ambiguity remains about the target or goal of the exercise in Ostrom's account. Are individual principles necessary to attain "institutional robustness" in the sense of arrangements that are long-enduring or capable of governing the activities of "generation after generation of appropriators," or to achieve sustainability evaluated in terms of the biogeophysical condition of the ecosystems in question?[2] No doubt it is reasonable to assume a close relationship between robustness and sustainability. An institutional arrangement will probably not survive over a long period of time if it fails to produce results that are sustainable in biogeophysical terms. Even so, for good reasons we must think carefully about the relationship between robustness and sustainability. If relevant ecosystems are subject to nonlinear changes that are natural or nonanthropogenic, achievement of robustness will require a well-developed capacity to monitor and adjust to natural changes in the state of key ecosystems. If such changes occasionally occur quite suddenly, creating a real danger of serious surprises, the pursuit of robustness will require creation and maintenance of early warning systems. And if substantial uncertainty exists about these matters, appropriators will be well advised to devote resources to improving their knowledge base regarding the behavior of the relevant ecosystems, introducing precautionary measures, or adopting a more complex strategy featuring both types of responses.

It is worth paying explicit attention as well to the fact that the design principles approach focuses on formulating necessary conditions or identifying links that are unconditional. From the perspective of demonstrating policy relevance, this feature has obvious attractions. It means, above all, that there should be no exceptions. If the existence of active and accountable monitors or the operation of credible sanctions is a true design principle, it would be foolhardy to ignore the prescriptive implications and suppose that it is possible in specific cases to cut corners when it comes to the organization of monitoring mechanisms capable of assessing CPR conditions or compliance mechanisms capable of imposing appropriate sanctions. But several other aspects of this feature are noteworthy as well. Design principles tell us nothing about sufficient conditions for achieving robustness or sustainability in managing human uses of CPRs. It is perfectly possible, therefore, that appropriators could adhere scrupulously to all the principles in their efforts to construct resource regimes but nonetheless fail to achieve robust or sustainable results. The emphasis on necessary conditions suggests also that available options to those desiring to achieve robust or sustainable outcomes may be quite limited, a conclusion that will seem counterintuitive both to analysts who think in terms of equifinality and to practitioners who believe that there is more than one way to solve most problems involving human actions (Young and Osherenko 1993). Coupled with the sheer number of distinct principles, the focus on necessity may engender a certain sense of pessimism as well. Ostrom identifies seven separate principles and suggests that we might ultimately have to add a few more.[3] Under the circumstances, policy makers may conclude understandably that they are being asked to make an enormous effort to devise arrangements that fulfill a long list of necessary conditions but that still offer no assurance that the results will prove to be robust or sustainable.[4]

Because design principles are framed as universal propositions, they should hold across all members of the relevant universe of cases. In Ostrom's work, this universe encompasses the set of small-scale CPRs, where small-scale is defined in terms of number of appropriators as well as spatial domain, and CPRs are construed as goods and services that are both nonexcludable and rival (or subtractable). Any situation with these characteristics in which appropriators succeed in producing results that are robust and sustainable without fulfilling one or another design principle would count as evidence against that principle. If this were to happen in more than the odd instance, a full-scale reassessment of the status of the relevant principle would be in order.

An obvious question is whether these design principles can be generalized to apply to cases that lie outside the universe of small-scale CPRs. In part, this is a matter of scale (Young 1994b). Can we scale up from small to large systems and ultimately to global systems in thinking about design principles relating to the institutional dimensions of environmental change (see chapter 6 for a more general account of the problem of scale)? In part, it is a matter of the nature of the problems at stake. To what extent do members of the overall set of environmental problems, including those occurring at the local level, exhibit the defining features of CPRs? It is understandable that Ostrom and colleagues have argued that there are reasonable grounds for arriving at optimistic answers to these questions (McGinnis and Ostrom 1996; Ostrom et al. 1999). But precisely because the incentive to do so is strong, we should examine the issues at stake with particular care.

Limits of Design Principles

The logic of design principles requires the existence of a universe of cases that is both well defined, at least in analytic terms, and homogeneous, at least with regard to factors relevant to the creation of regimes that are long enduring and capable of producing sustainable results. If we are unable to determine whether specific situations do or do not belong to the universe, the degree to which design principles constitute necessary conditions will be difficult to test. If, on the other hand, the boundaries of the universe are clear but the members of the universe are heterogeneous in important respects, individual design principles will be testable but they will fail to meet the test for (nearly) necessary conditions.

Is the universe of small-scale CPRs well defined and homogeneous? There is ample scope for debate regarding this question. To take a single example, it seems likely that those who think in terms of social-practice models of institutions would point to a number of factors (e.g., presence or absence of a strong sense of community) that could affect the way individual appropriators use living resources but that are not easily captured in the ideas of nonexcludability and subtractability (Singleton and Taylor 1992). But I do not wish to examine this issue in any depth here. For the sake of argument, let us accept the proposition that small-scale CPRs are, by and large, reasonably well defined and homogeneous. The relevant question becomes a matter of determining the extent to which the larger class of environmental problems arising at scales ranging from the microlevel of local ecosystems to the macrolevel of global ecosystems share the defining characteristics of small-scale CPRs. This is not a simple or straightforward question, a fact that should warn us against offering casual or offhand responses. Even so, I think it is fairly easy to show that the level of heterogeneity in the larger class of environmental problems is high enough to cast grave doubts on an assumption to the effect that what holds for small-scale CPRs will hold for other classes of problems involving environmental changes.

Consider several well-known analytic distinctions as vehicles for guiding our thinking about this issue. We begin with the familiar dichotomy between coordination problems and collaboration problems (Stein 1982). Whereas regimes that address collaboration problems (e.g., any situation that can be modeled as a prisoner's dilemma) leave individual participants with incentives to violate the rules or to cheat in meeting their commitments, coordination problems can be solved by developing regimes in which none of the participants experiences an incentive to cheat or defect.

In game-theoretic terms, regimes that deal with coordination problems are characterized by a stable equilibrium. Although it may be tempting to dismiss this finding as an interesting artifact with little relevance to the real world, evolution of social conventions or rules that leave everyone more or less content in a wide range of situations suggests that the category of coordination problems is far from empty in empirical terms. Much the same can be said regarding situations in which significant conflicts of interest are coupled with two or more equilibrium outcomes (e.g., battle of the sexes). Regime formation in such situations may be characterized by hard bargaining in which each party uses a variety of tactics to persuade or compel the other(s) to settle for its preferred outcome. But once a bargain is struck, members of the resultant regime will experience no incentives to cheat (Schelling 1960). In these cases, robustness will not depend on meeting requirements of principles dealing with monitoring and graduated sanctions. It may not even be necessary to worry about clearly defined boundaries, as long as new members are willing to accept or accede to existing rules when they become players in a continuing game.

Similar conclusions arise from consideration of well-known economic constructs. Where the problem centers on persuading participants to contribute to the supply of a public good in contrast to refraining from excessive use of a CPR, for instance, the challenge is to induce individual participants to take action rather than persuade them to avoid acting in a manner that is detrimental to social welfare.[5] Monitoring is seldom a problem when it comes to identifying contributions to the supply of a public good (e.g., improved knowledge of the functioning of key ecosystems); contributors will be anxious to make their contributions known to others. Nor is it necessary to worry about social boundaries. Because (pure) public goods, unlike CPRs, are not subtractable or rival, additional members can join the group with no negative consequences in terms of benefits accruing to original members. To the extent that they are prepared to join in the group's cost-sharing mechanism, in fact, contributions of new members will ease the burdens on original members. On the other hand, it is only fair to acknowledge that eliminating free riders in the context of public-goods problems (i.e., actors seeking to enjoy the benefits of a public good while contributing nothing to its supply) may well require establishment of effective compliance mechanisms. As experience with efforts to implement many international environmental regimes makes clear, it is no easy task to induce individual actors to live up to financial commitments they make at the outset, even when it is not difficult to monitor the extent to which they make contributions to the supply of a public good.[6] In some cases, parties cannot even bring themselves to file routine status reports in a timely manner.

Another common problem type arises in highly asymmetrical situations where the actions of one party (e.g., emissions of airborne pollutants, diversions of upstream waters into irrigation canals) produce externalities detrimental to the welfare of one or more of its neighbors. The problem here is to find ways to induce those who are sources of negative externalities to pay attention to the effects of their actions on the welfare of others. Although sanctions may be relevant, monitoring is seldom a problem since victims have well-developed incentives to prove that harm is being done to them and to provide evidence that will satisfy unbiased observers. Even more relevant is the observation that parties involved in such asymmetrical situations may well experience incentives to expand the issues included in problem sets or in negotiations aimed at resolving their differences (Sebenius 1983). If this is done in such a way that asymmetries relating to individual issues included in a larger package are matched with each other so that the overall bundle of issues becomes (roughly) symmetrical, what was initially a more or less severe conflict of interest can be transformed into a coordination problem in the sense that the benefits flowing to each of the participants are sufficiently large to suppress incentives they may have to cheat.

As this last observation suggests, the character of environmental problems is not wholly determined by nature. There is often room to influence the way a problem is framed for purposes of regime formation or institution building. This is especially true during agenda formation when issues under discussion are not crystallized through lengthy bargaining (Kingdon 1995; Young 1998). In some cases, this is a matter of spelling out the character of relatively specific issues (e.g., how should we think about the jurisdiction of coastal states in adjacent marine areas?). More often than not, however, it involves highly complex issues (e.g., should we include the full range of greenhouse gases within the framework of the climate regime?) or even the pros and cons of joining together several complex issues (e.g., ozone depletion, climate change, long-range air pollution) to build a regime. It does not require wholesale adoption of the tenets of social constructivism to see that often considerable room is available for developing different ways to think about complex environmental problems (Wendt 1992). Whereas the views of individual actors regarding such matters are often driven by prevailing conceptions of their own interests, substantial evidence indicates that cognitive forces that are not easy to reduce to simple expressions of actor interests play a role in defining issues and developing the discourses in terms of which they are addressed (Litfin 1994). This may lead to convergent characterizations of individual problems as cases of one or, at most, a few problem types. But evidence arising from actual social practices does not warrant a simple conclusion in this regard.

It is worth noting as well that efforts to determine what problem type a specific environmental concern exemplifies will often be afflicted by uncertainty either because the biogeophysical mechanisms at work are poorly understood or because of disagreements regarding either the extent to which the problem is anthropogenic or the nature of socioeconomic forces giving rise to the problem. In such cases, sharp disagreements about the characterization of problems are likely, coupled with a pronounced tendency to come to terms with uncertainty by selecting characterizations as much on the basis of self-interest as on the basis of objective or unbiased interpretations of available evidence. Settings of this sort often give rise to strong temptations to eliminate uncertainty by seeking to fit the full spectrum of environmental problems into a few familiar categories. Some writings on international environmental problems, for example, advance the claim that all such concerns are ultimately CPR problems, so that the challenge in each and every case is a matter of devising suitable arrangements to avoid the tragedy of the commons (Barkin and Shambaugh 1999). Comforting as this approach to uncertainty may be in analytic terms, however, evidence pertaining to actual cases does not support such a claim. Problems vary, opportunities for devising alternative characterizations of individual problems are substantial, and negotiations frequently begin in settings where no clear consensus exists among participants regarding proper characterization of the problem.

The conclusions I draw from this are that the larger universe of issues associated with environmental changes encompasses a range of problem types that are not reducible to a single homogeneous set, and that characterization of individual cases is by no means a routine—much less objective—process. Under the circumstances, we have no reason to expect a single set of design principles stated in the form of necessary conditions for achieving robustness or sustainability to stand up to vigorous testing. One size does not fit all when it comes to the creation of effective environmental regimes; design principles derived from a study of some members of the larger universe of problems run the risk of failing to produce the desired outcomes or leading to highly inefficient results when applied to others.

Does this mean that the search for design principles is a dead end and ought to be abandoned? Although the pitfalls associated with this approach are obviously severe, I believe it would be premature to reach such a conclusion. The development of design principles stated in the form of necessary conditions for success in problem solving is an attractive prospect, and we cannot rule out the possibility that additional research will turn up a set of such conditions that hold across the entire universe of environmental problems or even sizable subsets of the universe. Even so, it seems clear that we will not be able to count on this approach during the foreseeable future as a comprehensive method for bringing knowledge to bear on solving a variety of problems. Does the thesis of the previous section, then, imply that we should embrace the opposite view, asserting that each environmental problem is unique in a manner that leaves us with no alternative to treating every case as a universe of one? Although the reasoning underlying this reaction is easy enough to comprehend, I do not find such a conclusion any more satisfactory as a basis on which to initiate a constructive dialogue between analysts and practitioners who share an interest in solving a variety of more or less pressing environmental problems. What is necessary, at least at this stage, is an intermediate approach that avoids both pitfalls of excessive generalization and limitations arising from the treatment of each environmental problem as unique.

Institutional Diagnostics

The alternative I propose may be described as institutional diagnostics. The defining feature of this approach is an effort to identify important features of issues arising from environmental changes that can be understood as diagnostic conditions, coupled with an analysis of the design implications of each of these conditions. It is useful to treat this approach as an exercise in midrange generalization, coupled with liberal use of ceteris paribus assumptions. Instead of attempting to identify a given problem as a generic problem type (e.g., CPR) and then applying conditions regarded as necessary to successful treatment of all instances of that type, the diagnostic approach attempts to disaggregate environmental issues, identifying elements of individual problems that are significant from a problem-solving perspective and reaching conclusions about design features necessary to address each element.[7] This approach does not yield design principles or, for that matter, any generalizations that apply across the entire universe of environmental problems. Rather, it establishes a procedure in which the problems are considered on a case-by-case basis and prescriptions or recommendations are developed that take into account particular combinations of conditions. To borrow a medical term, this procedure may lead to identification of syndromes or recurrent combinations of diagnostic conditions that require similar treatment. But in many cases, the particular combination of conditions will be uncommon or even specific to the problem. As long as elements of these combinations are recognizable conditions that have identifiable implications for the design of management regimes, however, analysts and practitioners will be able to formulate recommendations for treatment of specific cases that are based on application of midrange generalizations.[8]

Beyond this, we must recognize at once a distinction between simple and complex diagnostics. Simple diagnostics involves a process in which the conditions associated with a specific environmental problem are examined one at a time to determine individual design implications. The implicit assumption is that individual diagnostic conditions do not interact to any significant degree. This assumption may be appropriate in some cases, at least as a first approximation or as a rough-and-ready guide. When this is so, simple diagnostics can prove sufficient for formulating recommendations pertaining to institutional design. But it is easy to see that in some cases individual diagnostic conditions will interact in ways that are too significant to ignore. When this occurs, it is essential to supplement simple diagnostics with some form of complex diagnostics.

Simple Diagnostics

The core of simple diagnostics is a set of three linked procedures: identifying a range of diagnostic conditions, evaluating the design implications associated with each condition, and developing interpretive skills necessary to apply these practices to specific cases. Informal efforts to make use of such procedures have long been a feature of exercises in regime formation at all levels of social organization. But the idea of developing institutional diagnostics into a self-conscious and more systematic approach to the design of environmental or resource regimes is largely unfamiliar. There is little prospect that we can devise a typology of diagnostic conditions that is exhaustive. As in most other diagnostic endeavors, the possibility of encountering conditions associated with specific environmental problems that are unfamiliar but turn out to have important prescriptive implications is always present. Even at this early stage, however, it is possible to give a substantial account of simple diagnostics as a procedure for addressing the institutional dimensions of environmental change.

Despite problems facing efforts to construct typologies, it is easy to identify a number of distinct diagnostic conditions that have significant implications for (re)designing institutions. Thus, table 7.2 encompasses three major classes of diagnostic conditions: ecosystem properties, actor attributes, and implementation issues. Ecosystem properties are features of relevant biogeophysical systems (and of our knowledge about them) that have important consequences for institutional design. Actor attributes are characteristics of the set of actors whose behavior gives rise to environmental problems (or who are engaged in efforts to deal with these problems) that have to be taken into account in designing arrangements to solve or ameliorate these problems. Implementation issues relate to fulfillment of institutional commitments that are important to the performance of arrangements created to address specific environmental problems. A brief discussion of each of these classes of conditions will flesh out the idea of simple diagnostics.

Table 7.2
Simple institutional diagnostics

Diagnostic condition	Design implication
Ecosystem properties	
Nonlinear or chaotic systems, surprises	Early warning devices/rapid response capability
Problem duration	Management structures
Functional interplay	Coordination mechanisms
Uncertainty, imperfect knowledge	Social learning, adaptability, precautionary approach
Actor attributes	
Variability of political and socioeconomic systems	Flexibility
Heterogeneity of member interests	Issue linkages
Asymmetries in causal responsibility	Emphasis on equity
Asymmetries in capacity	Capacity building, technology transfers
Implementation issues	
Violation tolerance	Systems of implementation review (SIRs)
Incentives to cheat	Sanctions, deterrence
Lack of transparency	Monitoring procedures
Malleability of rules	Certification, authoritative interpretation

Some systems are chaotic in the sense that they are subject to nonlinear and often poorly understood changes that can occur suddenly and produce major surprises (Ludwig, Hilborn, and Walters 1993). Those who regard climate change as particularly serious, for instance, typically assume that Earth's climate system is chaotic in this sense and point to indications in the paleoclimatic record of remarkably rapid and far-reaching nonanthropogenic changes occurring at various times in the past. The institutional implications of this condition are straightforward. To the extent that relevant biogeophysical systems are subject to chaotic behavior, it is important to devise effective early warning systems and to build rapid response capabilities into the design of institutional arrangements. The significance of this finding is particularly important in settings such as contemporary international society, where prevailing constitutive arrangements tend to impede efforts to respond quickly and decisively to changing circumstances.

The design implications of problem duration and functional interdependence are quite different. Whereas some problems can be solved once and for all, others involve activities that are likely to continue indefinitely. It is probable, for example, that the problem of ozone depletion will be solved by phasing out production and consumption of ozone-depleting substances, with the result that the ozone regime will work its way out of a job.[9] In contrast, a regime designed to govern human harvesting of renewable resources must be based on the expectation that its services will be required indefinitely. This distinction yields the following design implication. The longer the duration of the problem, the more sense it makes to invest resources in developing sophisticated administrative arrangements, funding mechanisms, and dispute-resolution procedures. Ad hoc arrangements that seem perfectly adequate to handle short-term needs will seem seriously deficient as permanent arrangements. For its part, functional interdependence is a matter of the extent to which problems dealt with under terms of separate regimes interact with one another. It is not necessary to be concerned about links between regimes in cases, such as management of fish stocks in the Bering and Barents Seas, where relevant biogeophysical systems are unrelated. But in cases such as ozone depletion and climate change, where functional interdependencies among distinct problems are substantial, it would be foolish not to think carefully about establishing institutional links or coordination mechanisms (including nesting and agreed-upon divisions of labor) in building regimes (Aggarwal 1998). In extreme cases, it may make sense to consider merging distinct institutional arrangements in the interests of coming to terms with the implications of functional interdependencies.

A related although somewhat different diagnostic condition involves the state of knowledge about ecosystem properties. In some cases, knowledge of relevant systems is highly sophisticated and the likelihood of surprises is low, although recent experience with efforts to manage marine fisheries should warn us against a tendency to become complacent in this regard (Harris 1998; Dobbs 2000). But in other cases, such as climate change, it is apparent that the biogeophysical systems are unusually complex, understanding of their behavior is limited, and the probability that unexpected but highly significant events will occurr is high (Bolin 1997). Of course, one response is to allocate substantial resources to improving knowledge regarding the behavior of the relevant systems; but this is likely to yield long-term improvement in management capabilities at best. In the meantime, one significant design implication arising from this condition features adoption of a precautionary approach and, as a result, establishment of relatively large margins of safety to avoid triggering unintended but irreversible changes in important systems.

Turning to actor attributes, another set of diagnostic conditions and design implications comes into focus. When actors involved in a problem—whether they are nation states or various types of nonstate actors—exhibit a high degree of heterogeneity with regard to their internal economic and political systems, it will make sense to create regimes that couple the rules and procedures required to solve problems with a high degree of latitude concerning procedures individual members are authorized to use in implementing these arrangements. Efforts to guide the behavior of rational, self-interested individuals through appeals to legitimate authority in contrast to measures that alter incentives are unlikely to succeed. But by the same token, incentive mechanisms may have little impact on the actions of those whose behavior is governed by the logic of appropriateness (March and Olsen 1998). Consider climate change. A successful regime in this area must be able to guide the actions of more than 180 countries that vary greatly in terms of the character of their political systems, nature of their economic systems, and stage of development. As a result, efforts to mandate uniformity with regard to policy instruments that member states employ to implement the terms of the regime within their domestic jurisdictions will almost certainly end in failure.

Heterogeneity with respect to the interests of relevant actors is another matter. Some observers believe that heterogeneity is a recipe for conflict that will make problems more difficult to solve. But this is not always the case. In some situations, heterogeneity regarding actor interests actually increases the scope for striking institutional bargains that satisfy the essential interests of all parties concerned. The fundamental design implication concerns the benefits of expanding the scope of institutional arrangements by adding issues and opening up opportunities to devise package deals in which individual actors make concessions on issues that are of lesser importance to them in return for concessions on the part of others regarding their priority concerns (Sebenius 1983). This process is well known in domestic settings where it is often described as log rolling or legislative bargaining, and sometimes produces packages of institutional arrangements that are more responsive to the parochial interests of individual actors than to the pursuit of common or societal goals such as sustainability. The process is less familiar at the international level, where individual issues are normally dealt with in separate forums. But as made clear by the recent effort to devise a package of provisions relating to climate change that encompasses targets and timetables on the one hand and policy instruments including joint implementation and emissions trading on the other, opportunities for devising package deals exist at the international level as well.

Asymmetries in causal responsibility and variations in capacity to implement institutional arrangements are additional conditions involving actor attributes that have significant design implications. When some actors bear the major burden of responsibility for causing environmental problems while others are likely to be most significantly affected by their consequences, success in creating effective institutions turns on issues relating to equity. Here again the case of climate change is instructive. So long as the United States, currently the source of approximately one-fourth of all greenhouse gas emissions, is either unable or unwilling to accept responsibility for its role in climate change, especially with regard to potential impacts on low-lying or small-island states that could be damaged severely by rising sea levels, the possibility of creating an effective climate regime is diminished.[10] One interesting response to this centers on asymmetries in the capacity of member countries to handle the burden of coping with climate change. Thus, the United States is not only the largest source of greenhouse gas emissions, it also has the greatest capacity of any member of international society to pay for various forms of mitigation and adaptation. One way to address such an equity problem is to devise an arrangement under which those with the greatest capacity to address the problem help those who are less fortunate by developing mechanisms featuring technology transfers and capacity building. The rationale underlying the clean development mechanism in the case of climate change reflects this way of thinking. Whether or not this approach proves successful in this case, the generic design implications are clear. Asymmetries in causal responsibility and in the ability to cope with problems must be recognized in design of arrangements.

Diagnostic conditions involving implementation issues are somewhat different from ecosystem properties and actor attributes. They deal with matters of institutional performance in contrast to features of relevant biogeophysical systems or attributes of actors. But these concerns can be approached in terms of institutional diagnostics as well. Regimes vary considerably in terms of both violation tolerance and incentives to cheat. Some regimes will unravel or collapse in the face of serious violations or even unsubstantiated allegations regarding violations (e.g., many arms control arrangements). Others can withstand relatively frequent violations without having their effectiveness called into question. In the case of climate change, aggregate trends in levels of greenhouse gas emissions count, rather than compliance of individual actors under specific circumstances. What this means is that the lower the level of violation tolerance associated with a regime, the greater will be the need for systems of implementation review capable of monitoring the behavior of individual members in some detail (Victor, Raustiala, and Skolnikoff 1998). Similar observations apply to incentives to cheat. It is unrealistic to hope that most major environmental concerns will take the form of coordination problems (Stein 1982). Yet problems vary substantially with regard to incentives to cheat, and especially the strength of incentives to cheat relative to gains derived from membership in the regime. Where incentives to cheat are comparatively weak, it will not be necessary to devote a great deal of time and energy to developing credible sanction procedures. The stronger the incentives to cheat, on the other hand, the greater the importance of sanctions that individual actors find credible (Downs, Rocke, and Barsoom 1996).

Much the same can be said regarding behavioral transparency and rule type. The degree to which the behavior of relevant actors is transparent is subject both to natural variation and to variation attributable to regime design. In some cases (e.g., clear-cutting of forests) it is difficult to engage in behavior leading to environmental problems in a covert or clandestine manner. Monitoring is not a major problem in such situations. More interesting, however, are cases in which behavioral transparency is naturally low but subject to manipulation on the part of those in a position to influence institutional design. Consider intentional oil pollution at sea and ozone depletion. Behavioral transparency in the case of oil pollution increased dramatically after a switch from rules based on discharge standards to rules based on equipment standards (Mitchell 1994). In the case of ozone depletion, it quickly became apparent that transparency would be much higher with regard to the actions of a handful of producers in contrast to millions of consumers, even though consumption of ozone-depleting substances is the ultimate concern (Benedick 1998). This suggests not only that the importance of investing in monitoring systems is a function of the degree to which relevant behavior is naturally transparent, but also that there is often considerable scope for designing rules in such a way as to increase transparency. In extreme cases, rules can be framed in a manner that places the burden of proof wholly on the subject. That is, subjects may be required to prove compliance rather than being treated as compliant unless and until some credible evidence of violations comes to hand. Short of this, however, it is often possible to frame the principal rules of regimes in different ways (Young 1999b). The value of investing resources in framing rules will vary as a function of other compliance concerns, including violation tolerance and incentives to cheat. But when rules are malleable, much is to be said for thinking carefully about the pros and cons of different ways of framing them.

What can we conclude from this account of simple diagnostics? The basic logic is clear. By examining diagnostic conditions one at a time, it is possible to think clearly about their implications for institutional design. Considerable variance exists among actual environmental issues in terms of the extent to which they exhibit these conditions. As a result, we cannot expect this approach to yield generalizations of the sort associated with design principles. What is more, this procedure resembles diagnostic procedures in other fields in the sense that using it to reach conclusions about specific problems requires considerable skill. It is perhaps inappropriate or unfair to describe diagnostics as an art rather than a science. But it is clear that those who use such procedures vary greatly in terms of their ability to arrive at accurate results. Whereas the unskilled diagnostician examines relevant conditions in a mechanical fashion, the truly talented diagnostician acquires an almost intuitive feel for the significance of key diagnostic conditions in specific situations. Of course, the importance of this skill depends on the complexity of the situation. Almost any well-trained person (or even a well-designed machine) can diagnose problems that are simple and straightforward. Real skill is required to deal with complex problems. This is one reason why climate change is so intriguing to students of environmental change. Not only is it likely to have profound consequences for human welfare, but it is perhaps the most complex environmental issue we have faced to date from the perspective of institutional diagnostics.

Complex Diagnostics

The idea of simple diagnostics is appealing precisely because it offers a procedure for thinking about the links between analysis and praxis with regard to institutional design that is relatively easy to understand. Yet it is based on the powerful assumption that individual diagnostic conditions do not interact with one another in significant ways. This assumption may be perfectly reasonable under some circumstances, but it is clear that interactions among relevant conditions are too important to ignore in other cases. When this happens, there is no alternative to supplementing simple diagnostics with complex diagnostics. The range of interactions among two or more diagnostic conditions is great; a useful typology of these interactions is currently beyond our grasp. Nonetheless, it is not hard to point to concrete cases that provide clear illustrations of interactions between diagnostic conditions.

One prominent example centers on interactions between nonlinear or chaotic systems and uncertainty or imperfect knowledge regarding the behavior of such systems. The origins of these interactions are apparent. Nonlinear systems, especially those in which surprises are common and where changes of state can occur quite suddenly, are far more difficult to understand than systems featuring stable equilibria. Not only are the systems themselves unusually complex, analytic tools necessary to understand them are comparatively weak. An obvious case is Earth's climate system. This poorly understood system is subject to nonlinear and sudden, surprising changes that may have profound consequences for human welfare. Yet our capacity to forecast, much less to predict, the occurrence of these changes is limited. What does this mean for the design of the climate regime? One response is to downplay the problem on the grounds that most societies have more pressing issues to contend with and that humans will learn their way out of this problem, as they have done with others, when the need arises. But an alternative response that may prove more convincing is to apply the precautionary principle, which, in this context, would mean including a margin of safety in setting targets and timetables in the regime. The argument here rests on the usual case for insurance; it makes sense to pay a small albeit significant price now to minimize the impact of potentially catastrophic occurrences at some later time.

A different illustration involves interactions between problem duration and efforts to come to terms with issues such as incentives to cheat and behavioral transparency. In dealing with problems that can be solved within a relatively short time, it is appropriate to limit investments in compliance mechanisms (e.g., programs designed to socialize key actors over the long term) that are costly and make sense only when they can be amortized over a long period of time. Thus, it seems reasonable to draw a distinction in these terms between the problem of ozone depletion on the one hand and the problems of climate change and loss of biological diversity on the other. Handled properly, phasing out the production and consumption of ozone-depleting substances will be complete within a decade or two at most. Once solved, it is not likely that this problem will reemerge in another form.[11] As a result, it is sensible to devise ad hoc arrangements, such as the Montreal Protocol Multilateral Fund, to persuade actors to comply with rules pertaining to the production and consumption of ozone-depleting substances and to avoid creating costly arrangements that may take on lives of their own and stay in business long after the problem is solved. Contrast this with climate change and biological diversity. These problems will be with us over the long term. The struggle to protect biological diversity, in particular, is destined to become a permanent item on the environmental agenda. The implications of this for institutional design are clear. Although initial costs may be high, it makes sense to make substantial investments early on in efforts to alter entrenched rights (e.g., traditional rights of private property owners) and to socialize actors into complying with regulatory rules as a matter of appropriate behavior rather than as a consequence of calculations of benefits and costs.[12]

When rules are malleable but violation tolerance is low and incentives to cheat are a real concern, another pattern of interaction surfaces. Different ways of framing key rules can affect prospects for eliciting compliance, in some cases dramatically. Where rules require large numbers of actors to refrain from behavior that is not naturally transparent, for instance, either the costs of obtaining compliance or the frequency of violations will rise sharply. Imagine the difficulties associated with persuading or compelling millions of consumers to comply with rules regarding consumption of products manufactured by processes involving substantial emissions of greenhouse gases. Eco-labeling will lead some consumers to alter their behavior at the margin in this connection. But such initiatives are hardly likely to solve the problem. In a case like this, much can be said for directing attention to alternative ways of framing the rules. One interesting option is to shift the burden of compliance from large numbers of consumers to much smaller numbers of producers. Requiring automobile manufacturers to meet well-defined fuel economy standards is a familiar case in point. A complementary strategy is to place the burden of demonstrating compliance on subjects, rather than assuming that the actions of subjects conform to the rules unless and until they prove to be noncompliant. These are strong measures, and obviously it is important to proceed with care in framing rules in specific cases. But it is clear that when concerns about violation tolerance and incentives to cheat are substantial yet rules are malleable, those concerned with problem solving will want to pay close attention to the pros and cons of alternative ways of formulating the rules.

These intuitively appealing examples can do no more than illustrate complex diagnostics. The range of potential interactions, including those involving three or more diagnostic conditions, is great. In analytic terms, this leads to an interesting trade-off. Procedures associated with simple diagnostics are much easier to use. But they may yield only prescriptions that are either obvious even to those who do not use systematic diagnostic procedures or inappropriate. Efforts to engage in complex diagnostics, in contrast, are more likely to yield counterintuitive results, but they may also lead to convoluted assessments whose design implications are both vague and difficult to explain. Fortunately, it is not necessary to make a definitive choice between simple and complex diagnostics. Of course, the fact that time and energy are finite will force analysts and practitioners alike to make trade-offs in dealing with specific cases. But no conceptual or analytic barriers discourage adoption of a mixed strategy that features the use of both forms of institutional diagnostics at the same time.

Common Diagnostic Errors

Like most other diagnostic endeavors, institutional diagnostics may yield erroneous conclusions in specific cases. Partly, this is a matter of the skill of the diagnostician. It does not require great insight to recognize considerable variation among practitioners with regard to their diagnostic skills. In part, however, diagnostic failures are products of standard errors that can afflict the efforts of even the most skillful diagnostician. We are not now in a position to provide a comprehensive list of these common errors; it may never be possible to produce a definitive list. But consider the following as illustrations: missing data, inappropriate models, unidentified interactions, and negotiated assessments.

Missing data are an obvious limitation, especially in efforts to assess ecosystem properties. Are relevant systems prone to nonlinear changes? If so, are shifts from one state to another likely to occur suddenly, as envisioned in the idea of punctuated equilibria, or more gradually, as envisioned in most discussions of evolutionary change? When large numbers of cases can be examined empirically or even subjected to controlled experiments, it is comparatively easy to answer these questions. But in cases such as Earth's climate system, involving biogeophysical systems that are large, complex, and unique, data necessary to evaluate important diagnostic conditions are likely to be few and far between. Among efforts regularly running into problems of this sort are attempts to determine whether the trends of the last few decades in global mean temperatures at Earth's surface are merely short-term cycles or early stages in a longer-term pattern of global warming. In the absence of a well-tested, predictive model of climate change, the natural reaction is to compare recent trends with a number of past changes. Although such efforts have yielded some suggestive results, the sparseness of data on past cycles places severe limits on diagnostic efforts of this sort.

A somewhat more insidious diagnostic error arises from the influence of models or analytic constructs that acquire a high level of credence but that ultimately prove to be inappropriate or applicable only to a limited set of cases. A classic example involves development and application of single-species, logistic models to assess the status of fish stocks and to prescribe regulatory procedures intended to produce maximum sustained yields (MSY) in individual fisheries over indefinite periods of time (Larkin 1977). In fact, large marine ecosystems feature complex interactions among different species and regularly undergo nonlinear changes after significant disturbances, such as depletion of species that make up sizable segments of the biomass (Sherman 1992; Wilson et al. 1994; National Research Council 1996). The use of MSY models to diagnose the condition of stocks and to make decisions about allowable harvest levels under such conditions frequently leads to undesirable and, in some cases, catastrophic results (McGoodwin 1990). Yet the use of such models dies hard. Not only are analysts likely to hold onto such simple models unless and until better alternatives become available, the models often make their way into the received and largely unquestioned wisdom of the relevant community. As the case of marine fisheries makes clear, this can lead to a disturbing record of erroneous diagnoses even among otherwise highly sophisticated observers.

With respect to unidentified interactions, a natural tendency is to think about environmental problems in terms of events occurring in stand-alone systems. Given the complexity of the dynamics of most human-dominated ecosystems, this tendency is certainly understandable. Yet it is increasingly clear that interactions among these systems are often significant and sometimes critical. Both ozone-depleting substances and many substitutes for them are greenhouse gases (Oberthür 1999). The fate of Earth's forests will have critical consequences both for climate change and maintenance of biological diversity. Shifts in the composition of species in large biogeophysical systems can accelerate or retard desertification or soil erosion. A successful effort to cope with environmental problems must rest on assumptions about the boundaries of relevant systems and the significance of interactions among distinct systems. Appealing as it is in conceptual terms, the familiar ecological principle that everything is related to everything else (Commoner 1972) does not offer much practical guidance for those seeking to solve specific problems. Yet failure to take into consideration the consequences of functional interdependencies is a common source of diagnostic errors.

Beyond this lies the problem of negotiated assessments. In the absence of decisive evidence regarding the status of biogeophyiscal systems and the effects of anthropogenic drivers, diagnoses are often subject to political processes. Even periodic assessments provided by the Intergovernmental Panel on Climate Change (IPCC), a body that has been praised deservedly for the rigor and unbiased character of its procedures, are ultimately negotiated in a setting that is by no means strictly scientific (Edwards and Schneider 2001). In one sense, one cannot escape from such processes. All responses to environmental problems, including decisions to do nothing, rest on assessments regarding the behavior of the relevant systems. All assessments, regardless of the care with which they are carried out, involve judgments that can never be strictly objective or unaffected by the beliefs and values of those who make them (Miller and Edwards 2001). Even so, one sees a striking difference between assessments that are based on careful and transparent procedures and those that are products of obscure bargaining among obviously self-interested actors. The contrast between the work of the IPCC or the International Council for the Exploration of the Sea and the activities of scientific committees attached to many regimes dealing with marine living resources in these terms is striking. Institutional diagnoses cannot escape the need for judgment. The important point is to avoid setting too much store by negotiated assessments.

In the nature of things, institutional diagnosis is fallible. The role of judgment is substantial, and absence of simple formulas places a premium on contributions of the skilled diagnostician. There is no surefire way to avoid all diagnostic errors. But one way to minimize the frequency and impact of such pitfalls is to identify major categories of these errors as clearly as possible and to encourage those engaged in institutional diagnostics to consider the relevant dangers carefully in dealing with specific cases. Awareness of shortcomings of MSY models, for instance, does not guarantee success in designing successful regimes for marine living resources. But it may well play some role in allowing those responsible for the creation and implementation of fisheries regimes to avoid pitfalls that have occurred again and again in efforts to manage fisheries and, for that matter, many other human-dominated ecosystems (Vitousek et al. 1997).

Final Words

Some observers take the view that it is desirable to separate analysis and praxis in thinking about the institutional dimensions of environmental problems. The argument here turns on the proposition that it is desirable to insulate analysis from the more political processes involved in efforts to solve specific problems. The principal concern is that interactions between analysts and practitioners will corrupt efforts to produce usable knowledge and in the end lead to results that are both unsatisfactory in analytic terms and of little value in applied terms. Although this concern is real and understandable, I do not subscribe to the conclusion it suggests. Analytic processes themselves have a political dimension, regardless of our efforts to achieve objectivity. It is not necessary to adopt all the arguments often grouped under the heading of the social studies of science to acknowledge this fact (Jasanoff and Wynne 1998). But even more important, in my judgment, is the fact that continuing dialogue between analysts and practitioners can prove beneficial to both communities, helping analysts to test existing ideas and generate new insights, and assisting practitioners to broaden their thinking and to avoid simplistic analogies or inappropriate responses to specific problems. Both development of design principles and pursuit of institutional diagnostics can serve as vehicles for fostering such a dialogue; we do not have to make a choice between these approaches to the production of usable knowledge. Nonetheless, I am convinced that institutional diagnostics is an approach that is particularly well suited to facilitating productive exchanges between analysts and practitioners and that deserves increased attention among those desiring to generate usable knowledge pertaining to the institutional dimensions of environmental change.

Notes

Chapter 1

1. Formally, this is the 1995 Agreement for the Implementation of the 1982 United Nations Convention on the Law of the Sea, Relating to the Conservation and Management of Straddling Fish Stocks and Highly Migratory Fish Stocks.

2. The terminology in the table differs somewhat from the parallel terminology in the IDGEC Science Plan. Specifically, I have embellished the original functional-political distinction to emphasize that functional linkages arise from biogeophysical and socioeconomic interdependencies, whereas political linkages involve deliberate efforts to engage in institutional design or management.

Chapter 2

1. This point can be framed in terms of "the notion that actors and their interests are institutionally constructed" (Powell and DiMaggio 1991, 28).

2. Common examples are charges levied on certain types of behavior and income derived from the sale of tradable permits.

3. Of course, it is possible to construct rational-choice models of interactions among interest groups. But much of the elegance and tractability associated with unitary-actor models is lost in the process.

4. Where compliance is inexpensive or even cost free, models based on utilitarian assumptions will predict that violations will occur infrequently, if at all.

Chapter 3

1. Set forth in the Homestead Act of 1862, this system allowed settlers to gain title to land included in the public domain largely on the basis of their own labor.

2. Advances in technology occurring in recent decades have led to significant improvements in monitoring whale stocks. But even now, significant uncertainties obtain regarding the status of many stocks.

3. Pareto optimality occurs when no feasible change from the status quo will enhance the welfare of some or all members of a group without leaving any member worse off than before.

4. The RMP itself calls for a high degree of caution. But this has not stopped those opposed to harvesting whales from opposing its implementation on precautionary grounds.

Chapter 4

1. Put another way, these chapters place primary emphasis on the lower left and upper right compartments of table 1.1.

2. Interactions between national and state-provincial arrangements and between state-provincial and local arrangements are common in most domestic settings. So also are interactions between global and regional arrangements in international settings. The goal of this chapter, however, is to explore the nature of vertical interplay by analyzing several prominent examples rather than to present a taxonomy of various forms of vertical interplay.

3. Both Principle 21 of the 1972 Stockholm Declaration and Principle 2 of the 1992 Rio Declaration, for instance, declare that "States have . . . the sovereign right to exploit their own resources . . ."

4. Recent settlements of comprehensive claims with aboriginal peoples in the Canadian North have reduced the scope of public property somewhat and at the same time introduced some interesting arrangements featuring more complex systems of land tenure. Even so, public land ownership remains the norm in Canada.

5. In the Alaska Native Claims Settlement Act of 1971 (PL 92-203), for instance, the federal government awarded full title to almost 44 million acres of land to Native corporations, but at the same time declared, "All aboriginal titles, if any, and claims of aboriginal title in Alaska . . . including any aboriginal hunting or fishing rights that may exist, are hereby extinguished" (Sec. 4b).

6. For an early, but still helpful treatment of property systems as social institutions, see Hallowell (1943).

7. In cases where traditional socioeconomic practices have given way to mixed economies, local peoples may experience a growing need to exploit resources to generate cash flow.

8. For evidence of similar interactions occurring in other parts of the world, see Gibson et al. (2000).

9. For an extended account of the role of rules in use and the relationship between such rules and formal rules, see Ostrom (1990).

10. The doughnut hole consists of a pocket of high seas wholly surrounded by EEZs of Russia and the United States.

11. Some arrangements (e.g., the ozone and climate regimes) differentiate among classes of members with regard to specification of obligations and application of rules. This has given rise to what is known as the principle of common but differentiated responsibility.

12. See the *Review of Progress towards the Year 2000 Objective,* issued on November 5, 2000, by the International Tropical Timber Organization as document ITTC(XXVIII)/9/Rev.2.

13. Updates on the work of CAFF appear regularly in the *Arctic Bulletin,* published four times a year under the auspices of the WWF Arctic Programme.

14. The Straddling Fish Stocks Agreement also encompasses other goals, such as devising rules designed to achieve sustainable harvesting of highly migratory species (e.g., tuna).

15. The long-standing tension in American wildlife law regarding allocation of authority between governments of the states and the federal government illustrates this debate.

Chapter 5

1. Framing issues with regard to large-scale environmental matters is a major focus of analysis in the Global Environmental Assessment project based at Harvard University (Social Learning Group 2001).

2. The North Sea regime has evolved into a more complex institutional system over time. Among other things, this has led to greater integration of efforts to deal with vessel-source and land-based pollution. But the regime remains a free-standing arrangement.

3. Many of the same individuals participated in the two sets of negotiations. But the fact that negotiations occurred under different auspices shaped the course of these processes of regime formation in significant ways.

4. This does not mean that they always or even usually succeed in these terms. The politics of regime formation are too complex to support simple generalizations of this sort.

5. Note that this generally involves a search for what might be called the highest common denominator rather than the lowest common denominator, as some commentators have suggested. For a clear account framed in terms of the idea of the least ambitious program, see Underdal (1980).

6. Note that these organizations may—but need not—be elements of the regimes themselves.

7. The funding mechanism for the desertification regime, in contrast, is the International Fund for Agricultural Development.

8. Some observers worry about the prospect of conflicts between the rulings of ITLOS and those of the International Court of Justice, and some states have rejected the jurisdiction of ITLOS in ratifying UNCLOS.

Chapter 6

1. In reality, some resources are likely to exhibit degrees of excludability or rivalness. For a general account of such matters described in terms of "fuzzy sets," see Ragin (2000).

2. Several caveats are in order in applying these analytic distinctions to real-world situations. Whereas some public goods can become degraded or congested as the size of the user group grows, others are strictly nonrival in the sense that they are not subject to degradation regardless of group size. Similarly, club goods may be subject to crowding or degradation if organizers-managers are unable or unwilling to construct or implement effective entry barriers.

3. Note that in the case of the electromagnetic spectrum, advances in technology may make this natural resource nonrival for all practical purposes.

4. Under some circumstances, regulatory rules or decision-making procedures may suffer from congestion in the sense that regimes typically have a limited capacity to deal with many issues at the same time. Note also that some members of the recipient group may place a negative value on specific public goods. Those who dislike a particular regime, for instance, may regard it as a public "bad," even though they cannot exclude themselves from its effects.

5. Ample evidence supports the view that power relations are a major determinant of outcomes at both local and international levels. But since the role of power is similar in the two settings, I do not deal with it at length in this discussion of scaling up.

6. For an argument that these norms are often violated in practice, see Krasner (1999).

7. Yet ambitious developments, such as the dispute resolution procedures of the World Trade Organization and ITLOS, indicate that international society is changing in these terms.

Chapter 7

1. This phrase is the subtitle of Ostrom 1990.

2. Ostrom herself tends to emphasize the idea of long-enduring arrangements, but the general literature on the tragedy of the commons focuses on the biogeophysical condition of CPRs or, more generally, on sustainability (Hardin and Baden 1977; Baden and Noonan 1998).

3. To these, she added an eighth condition relating specifically to CPRs that are parts of larger systems (Ostrom 1990, 90).

4. Of course, many long-enduring CPR institutions that Ostrom considers in her study have evolved over time without conscious efforts on the part of their creators. In such cases, the number of necessary conditions may be irrelevant. But so also is any activity we might reasonably describe as institutional design.

5. Whereas CPRs are goods that are nonexcludable and rival or subtractable, (pure) public goods are both nonexcludable and nonrival (Olson 1965).

6. In some cases an individual actor may place such a high value on a public good that it is willing to shoulder the cost of supplying the good without exerting pressure on other members of the group to make contributions. In the literature on public goods, that is known as a privileged group (Olson 1965). Among students of international relations, an actor that shoulders the full burden of supplying a public good is known as a hegemon (Keohane 1984; Snidal 1985).

7. Construed in this way, the practice of institutional diagnostics has something in common with Elster's (1999) account of mechanisms.

8. Note that this approach bears a distinct resemblance to the procedure that an ecologist asked to make recommendations for the management of a specific ecosystem (e.g., a woodlot) would employ.

9. The effects of ozone depletion will be felt for some time to come due to the fact that some ozone-depleting substances have long periods of residence in Earth's atmosphere. Once satisfactory alternatives are introduced, however, it is unlikely that producers and consumers will revert to the use of these substances.

10. An additional complication arises when actors (e.g., India in the case of climate change) that bear little responsibility for a problem in its current form are expected to become increasingly responsible for the problem in the future.

11. Of course, it is possible, even probable, that problems analogous to ozone depletion, in the sense that they are externalities of the use of chemicals developed for other purposes, will arise from time to time.

12. This argument may go some way toward justifying the expenditure of more time and energy on the organization of the Global Environment Facility than the Multilateral Fund.

References

Aggarwal, Vinod K., ed. 1998. *Institutional Designs for a Complex World: Bargaining, Linkages, and Nesting.* Ithaca: Cornell University Press.

Allison, Graham T. 1971. *The Essence of Decision.* Boston: Little, Brown.

Anand, R. P. 1983. *Origin and Development of the Law of the Sea: History of International Law Revisited.* The Hague: Martinus Nijhoff.

Apostle, Richard, Gene Barrett, Peter Holm, Svein Jentoft, Leigh Mazany, Bonnie McCay, and Knut Mikalsen. 1998. *Community, State, and Market on the North Atlantic Rim.* Toronto: University of Toronto Press.

Axelrod, Robert. 1984. *The Evolution of Cooperation.* New York: Basic Books.

Axelrod, Robert. 1997. *The Complexity of Cooperation: Agent-Based Models of Competition and Collaboration.* Princeton, NJ: Princeton University Press.

Baden, J. A. and D. S. Noonan, eds. 1998. *Managing the Commons.* Bloomington: Indiana University Press.

Baldwin, David, ed. 1993. *Neorealism and Neoliberalism: The Contemporary Debate.* New York: Columbia University Press.

Balton, David A. 2001. "The Bering Sea Doughnut Hole Convention: Regional Solution, Global Implications." In Olav Schram Stokke, ed., *Governing High Seas Fisheries: The Interplay of Global and Regional Regimes.* London: Oxford University Press, 143–177.

Barkin, J. Samuel and George E. Shambaugh, eds. 1999. *Anarchy and the Environment: The International Relations of Common Pool Resources.* Albany: SUNY Press.

Barrett, Scott. 1999. "A Theory of Full International Cooperation." *Journal of Theoretical Politics,* 11: 519–541.

Benedick, Richard. 1998. *Ozone Diplomacy: New Directions in Safeguarding the Planet.* Rev. ed. Cambridge: Harvard University Press.

Berkes, Fikret, ed. 1989. *Common Property Resources: Ecology and Community-Based Sustainable Development.* London: Belhaven.

Berkes, Fikret. 1999. *Sacred Ecology: Traditional Ecological Knowledge and Resource Management.* Philadelphia: Taylor & Francis.

Berkes, Fikret. 2000. "Cross-Scale Institutional Linkages: Perspectives from the Bottom Up." Paper prepared for the meeting of the International Association for the Study of Common Property, Bloomington, Indiana.

Berkes, Fikret and Carl Folke, eds. 1998. *Linking Social and Ecological Systems: Management Practices and Social Mechanisms for Building Resilience.* Cambridge: Cambridge University Press.

Biermann, Frank. 2000. "The Case for a World Environment Organization." *Environment,* 42(9): 22–31.

Bolin, Bert. 1997. "Scientific Assessment of Climate Change." In Gunnar Fermann ed., *International Politics of Climate Change: Key Issues and Critical Actors.* Oslo: Scandinavian University Press, 83–109.

Breitmeier, Helmut, Marc A. Levy, Oran R. Young, and Michael Zürn. 1996. "The International Regimes Database as a Tool for the Study of International Cooperation." International Institute for Applied Systems Analysis. IIASA WP-96-160.

Brennan, Geoffrey and James M. Buchanan. 1985. *The Reason of Rules: Constitutional Political Economy.* Cambridge: Cambridge University Press.

Bromley, Daniel W., ed. 1992. *Making the Commons Work: Theory, Practice, and Policy.* San Francisco: ICS Press.

Brubaker, Sterling, ed. 1984. *Rethinking the Federal Lands.* Washington, DC: Resources for the Future.

Buchanan, James M., R. D. Tollison, and Gordon Tullock, eds. 1980. *Towards a Theory of the Rent Seeking Society.* College Station: Texas A & M Press.

Bull, Hedley. 1977. *The Anarchical Society: A Study of Order in World Politics.* New York: Columbia University Press.

Burger, Joanna, Elinor Ostrom, Richard B. Norgaard, David Policansky, and Bernard D. Goldstein, eds. 2001. *Protecting the Commons: A Framework for Resource Management in the Americas.* Washington, DC: Island Press.

CAFF. 1996. "Circumpolar Protected Areas Network (CPA)—Strategy and Action Plan." *CAFF Habitat Conservation Reports no. 6.*

Caron, David D. 1995. "The International Whaling Commission and the North Atlantic Marine Mammals Commission: The Institutional Risks of Coercion in Consensual Structures," *American Journal of International Law,* 89: 154–174.

Chayes, Abram and Antonia Handler Chayes. 1995. *The New Sovereignty: Compliance with International Regulatory Agreements.* Cambridge: Harvard University Press.

Chertow, Marian R. and Daniel C. Esty. 1997. *Thinking Ecologically: The Next Generation of Environmental Policy.* New Haven, CT: Yale University Press.

Clarke, Jeanne Nienaber and Daniel C. McCool. 1996. *Staking out the Terrain: Power and Performance among Natural Resource Agencies.* Albany: SUNY Press.

Claude, Inis L., Jr. 1988. *States and the Global System: Politics, Law and Organization.* New York: St. Martin's Press.

Cleveland, Cutler, Robert Costanza, Thrainn Eggertsson, Louise Fortmann, Bobbi Low, Margaret McKean, Elinor Ostrom, James Wilson, and Oran R. Young. 1996. "A Framework for Modeling the Linkages between Ecosystems and Human Systems." Beijer discussion paper series no. 76. Stockholm: Beijer International Institute of Ecological Economics.

Commoner, Barry. 1972. *The Closing Circle*. New York: Bantam Books.

Conca, Ken. 1995. "Environmental Protection, International Norms, and State Sovereignty: The Case of the Brazilian Amazon." In Gene M. Lyons and Michael Mastanduno, eds., *Beyond Westphalia: State Sovereignty and International Intervention*. Baltimore: Johns Hopkins University Press, 147–169.

Conca, Ken. 2001. "Environmental Cooperation and International Peace." In Paul F. Diehl and Nils Petter Gleditch, eds., *Environmental Conflict*. Boulder, CO: Westview Press, 225–247.

Costanza, Robert and Carl Folke. 1996. "The Structure and Function of Ecological Systems in Relation to Property-Rights Regimes." In Susan Hanna, Carl Folke, and Karl-Göran Mäler, eds., *Rights to Nature*. Washington, DC: Island Press, 13–34.

Cutler, A. Claire, Virginia Haufler, and Tony Porter, eds. 1999. *Private Authority and International Affairs*. Albany: SUNY Press.

Dauvergne, Peter. 1997a. *Shadows in the Forest: Japan and the Politics of Timber in Southeast Asia*. Cambridge: MIT Press.

Dauvergne, Peter. 1997b. "A Model of Sustainable Trade in Tropical Timber." *International Environmental Affairs*, 9: 3–21.

Dessler, David. 1989. "What's at Stake in the Agent-Structure Debate?" *International Organization*, 43: 441–473.

Dobbs, David 2000. *The Great Gulf: Fishermen, Scientists, and the Struggle to Revive the World's Greatest Fishery*. Washington, DC: Island Press.

Downs, Anthony. 1972. "Up and Down with Ecology—The 'Issue-Attention Cycle.'" *Public Interest*, 28: 38–50.

Downs, George W., David M. Rocke, and Peter N. Barsoom. 1996. "Is the Good News about Compliance Good News about Cooperation?" *International Organization*, 50: 379–406.

Edwards, Paul N. and Stephen H. Schneider. 2001. "Self-Governance and Peer Review in Science-for-Policy: The Case of the IPCC." In Clark Miller and Paul N. Edwards, eds., *Changing the Atmosphere: Expert Knowledge and Global Environmental Governance*. Cambridge: MIT Press.

Elster, Jon. 1999. *Alchemies of the Mind*. Cambridge: Cambridge University Press.

Esty, Daniel C. 1993. *Greening the GATT*. Washington, DC: Institute for International Economics.

Fairman, David. 1996. "The Global Environment Facility: Haunted by the Shadow of the Future." In Robert O. Keohane and Marc A. Levy, eds., *Institutions for Environmental Aid*. Cambridge: MIT Press, 55–87.

Fienup-Riordan, Ann. 1990. *Eskimo Essays: Yu'pik Lives and How We See Them*. New Brunswick, NJ: Rutgers University Press.

Fondahl, Gail. 1998. *Gaining Ground: Evenkis, Land, and Reform in Southeastern Siberia*. Boston: Allyn & Bacon.

Franck, Thomas M. 1990. *The Power of Legitimacy among Nations*. New York: Oxford University Press.

French, Hilary F. 1997. "Learning from the Ozone Experience." In Lester R. Brown, Christopher Flavin, and Hilary F. French, eds., *State of the World 1997*. New York: W.W. Norton, 151–171.

Friedheim, Robert L. 1993. *Negotiating the New Ocean Regime*. Columbia: University of South Carolina Press.

Friedheim, Robert L., ed. 2001. *Toward a Sustainable Whaling Regime*. Seattle: University of Washington Press.

Gehring, Thomas. 1994. *Dynamic International Regimes: Institutions for International Environmental Governance*. Frankfurt am Main: Peter Lang.

Gibson, Clark. 1999. *Politicians and Poachers: The Political Economy of Wildlife Policy in Africa*. Cambridge: Cambridge University Press.

Gibson, Clark C., Margaret A. McKean, and Elinor Ostrom, eds. 2000. *People and Forests: Communities, Institutions, and Governance*. Cambridge: MIT Press.

Goldstein, Joshua. 1986. *Long Cycles in War and Economic Growth*. New Haven, CT: Yale University Press.

Golovnev, Andrei and Gail Osherenko. 1999. *Siberian Survival: The Nenets and Their Story*. Ithaca: Cornell University Press.

Gulland, J. A. 1974. *The Management of Marine Fisheries*. Seattle: University of Washington Press.

Guppy, Nicholas. 1996. "International Governance and Regimes Dealing with Land Resources from the Perspective of the North." In Oran R. Young, George J. Demko, and Kilaparti Ramakrishna, eds., *Global Environmental Change and International Governance*. Hanover, NH: University Press of New England, 136–162.

Haas, Peter M. and Ernst B. Haas. 1995. "Learning to Learn: International Governance." *Global Governance*, 1: 255–284.

Hallowell, A. Irving. 1943. "The Nature and Function of Property as a Social Institution." *Journal of Legal and Political Sociology*, 1: 115–138.

Hanna, Susan, Heather Blough, Richard Allen, Suzanne Iudicello, Gary Matlock, and Bonnie McCay. 2000. *Fishing Grounds: Defining a New Era in American Fisheries Management*. Washington, DC: Island Press.

Hardin, Garrett. 1968. "The Tragedy of the Commons." *Science*, 162: 1343–1348.

Hardin, Garrett and John Baden, eds. 1977. *Managing the Commons*. San Francisco: W.H. Freeman.

Hardin, Russell. 1982. *Collective Action.* Baltimore: Johns Hopkins University Press.

Harris, Michael. 1998. *Lament for an Ocean: The Collapse of the Atlantic Cod Fishery.* Toronto: McClelland & Stewart.

Hart, H. L. A. 1961. *The Concept of Law.* Oxford: Oxford University Press.

Hasenclever, Andreas, Peter Mayer, and Volker Rittberger. 1999. "Distributive Justice and the Robustness of International Regimes: A Preliminary Project Report." Paper prepared for the workshop on the Study of Regime Consequences, Oslo.

Hays, Samuel P. 1959. *Conservation and the Gospel of Efficiency: The Progressive Conservation Movement.* Cambridge: Harvard University Press.

Helm, Carsten and Detlev F. Sprinz. 1999. "Measuring the Effectiveness of International Environmental Regimes." *PIK Report no. 52.*

Herr, Richard, ed. 1995. *Antarctica Offshore: A Cacophony of Regimes?* Hobart: University of Tasmania.

Higgins, Rosalyn. 1994. *Problems and Process: International Law and How We Use It.* Oxford: Clarendon Press.

Hoel, Alf Håkon. 1993. "Regionalization of International Whale Management: The Case of the North Atlantic Marine Mammals Commission." *Arctic,* 46: 116–123.

Hoel, Alf Håkon. 1999. "Institutional Interplay among Environmental Institutions in the Arctic." Paper presented at the open meeting of the Human Dimensions of Global Environmental Change Community, Shonan Village, Japan.

Holling, C. S. and Steven Sanderson. 1996. "Dynamics of (Dis)harmony in Ecological and Social Systems." In Susan S. Hanna, Carl Folke, and Karl-Göran Mäler, eds., *Rights to Nature.* Washington, DC: Island Press, 57–85.

Hønneland, Geir. 2000. *Coercive and Discursive Compliance Mechanisms in the Management of Natural Resources: A Case Study of the Barents Sea Fisheries.* Dordrecht: Kluwer Academic Publishers.

Humphreys, David. 1996. "Hegemonic Ideology and the International Tropical Timber Organization." In John Vogler and Mark Imber, eds., *The Environment and International Relations.* London: Routledge, 215–233.

Huntington, Henry P. 1997. "The Arctic Environmental Protection Strategy and the Arctic Council: A Review of United States Participation and Suggestions for Future Involvement." Report prepared for the Marine Mammals Commission.

Hurrell, Andrew. 1992. "Brazil and the International Politics of Deforestation." In Andrew Hurrell and Benedict Kingsbury, eds., *The International Politics of the Environment.* Oxford: Oxford University Press, 398–429.

Hurrell, Andrew. 1993. "International Society and the Study of Regimes." In Volker Rittberger, ed., *Regime Theory and International Relations.* Oxford: Clarendon Press, 49–72.

Independent World Commission on the Oceans. 1998. *The Ocean Our Future.* Cambridge: Cambridge University Press.

Iudicello, Suzanne, Michael Weber, and Robert Wieland. 1999. *Fish, Markets, and Fishermen: The Economics of Overfishing.* Washington, DC: Island Press.

Jasanoff, Sheila and Bryan Wynne. 1998. "Science and Decisionmaking." In S. Rayner and E. Malone, eds., *Human Choice and Climate Change.* Vol. 1. *The Societal Framework.* Columbus, OH: Battelle Press, 1–87.

Jodha, Narpart S. 1996. "Property Rights and Development." In Susan S. Hanna, Carl Folke, and Karl-Göran Mäler, eds., *Rights to Nature.* Washington, DC: Island Press, 205–220.

Joyner, Christopher C. 1998. *Governing the Frozen Commons: The Antarctic Regime and Environmental Protection.* Columbia: University of South Carolina Press.

Juda, Lawrence. 1996. *International Law and Ocean Use Management: The Evolution of Ocean Governance.* London: Routledge.

Kaufman, Herbert. 1960. *The Forest Ranger: A Study in Administrative Behavior.* Baltimore: Johns Hopkins University Press.

Kennedy, Paul. 1987. *The Rise and Fall of the Great Powers: Economic Change and Military Conflict from 1500 to 2000.* New York: Random House.

Keohane, Robert O. 1984. *After Hegemony: Cooperation and Discord in the World Political Economy.* Princeton, NJ: Princeton University Press.

Keohane, Robert O. and Marc A. Levy, eds. 1996. *Institutions for Environmental Aid.* Cambridge: MIT Press.

King, Gary, Robert O. Keohane, and Sydney Verba. 1994. *Designing Social Inquiry: Scientific Inferences in Qualitative Research.* Princeton, NJ: Princeton University Press.

Kingdon, John W. 1995. *Agendas, Alternatives, and Public Policies,* 2nd ed. New York: Harper Collins.

Klyza, Christopher M. 1996. *Who Controls Public Lands?* Chapel Hill: University of North Carolina Press.

Krasner, Stephen D. 1991. "Global Communications and National Power: Life on the Pareto Frontier." *World Politics,* 43: 336–366.

Krasner, Stephen D. 1999. *Sovereignty: Organized Hypocrisy.* Princeton, NJ: Princeton University Press.

Kratochwil, Friedrich. 1989. *Rules, Norms, and Decisions: On the Conditions of Practical and Legal Reasoning in International Relations and Domestic Affairs.* Cambridge: Cambridge University Press.

Krech, Shepard, III. 1999. *The Ecological Indian: Myth and History.* New York: W.W. Norton.

Krupnik, Igor. 1993. *Arctic Adaptations: Native Whalers and Reindeer Herders of Northern Eurasia.* Hanover, NH: University Press of New England.

Larkin, P. A. 1977. "An Epitaph for the Concept of Maximum Sustained Yield." *Transactions of the American Fisheries Society,* 106: 1–11.

Levy, Marc A., Oran R. Young, and Michael Zürn. 1995. "The Study of International Regimes." *European Journal of International Relations,* 1: 267–330.

Lipschutz, Ronnie D. and Ken Conca, eds. 1993. *The State and Social Power in Global Environmental Politics.* New York: Columbia University Press.

Litfin, Karen T. 1994. *Ozone Discourses: Science and Politics in Global Environmental Cooperation.* New York: Columbia University Press.

Litfin, Karen T., ed. 1998. *The Greening of Sovereignty.* Cambridge: MIT Press.

Lowi, Miriam. 1995. *Water and Power: The Politics of a Scarce Resource in the Jordan River Basin.* Cambridge: Cambridge University Press.

Luce, R. Duncan and Howard Raiffa. 1957. *Games and Decisions.* New York: Wiley.

Ludwig, D., R. Hilborn, and C. Walters. 1993. "Uncertainty, Resource Exploitation, and Conservation: Lessons from History." *Science,* 260: 17, 36.

Lynge, Finn. 1992. *Arctic Wars, Animal Rights, Endangered Peoples.* Hanover, NH: University Press of New England.

Lyons, Geme M. and Michael Mastanduno, eds. 1995. *Beyond Westphalia: State Sovereignty and International Intervention.* Baltimore: Johns Hopkins University Press.

March, James G. and Johan P. Olsen. 1998. "The Institutional Dynamics of International Political Orders." *International Organization,* 52: 943–969.

McCay, Bonnie J. and James M. Acheson, eds. 1987. *The Question of the Commons: The Culture and Ecology of Communal Resources.* Tucson: University of Arizona Press.

McGinnis, Michael and Elinor Ostrom. 1996. "Design Principles for Local and Global Commons." In Oran R. Young, ed., *The International Political Economy and International Institutions.* Vol. 2. Cheltenham, UK: Edward Elgar, 465–493.

McGoodwin, James. 1990. *Crisis in the World's Fisheries: People, Problems, and Policies.* Stanford, CA: Stanford University Press.

McPhee, John. 1971. *Encounters with the Archdruid.* New York: Farrar, Straus, & Giroux.

Mearsheimer, John J. 1994–1995. "The False Promise of International Institutions." *International Security,* 19: 5–49.

Meyer, John W., David John Frank, Ann Hironaka, Evan Schofer, and Nancy Brandon Tuma. 1997. "The Structuring of a World Environmental Regime, 1870–1990." *International Organization,* 51(4): 623–651.

Miles, Edward L., Arild Underdal, Steinar Andresen, Jørgen Wettestad, Jon Birger Skjærseth, and Elaine M. Carlin. 2001. *Explaining Regime Effectiveness: Confronting Theory with Evidence.* Cambridge: MIT Press.

Miller, Clark and Paul N. Edwards, eds. 2001. *Changing the Atmosphere: Expert Knowledge and Environmental Governance.* Cambridge: MIT Press.

Mitchell, Ronald B. 1994. *Intentional Oil Pollution at Sea: Environmental Policy and Treaty Compliance.* Cambridge: MIT Press.

Mitchell, Ronald B. 1995. "Compliance with International Treaties: Lessons from Intentional Oil Pollution." *Environment,* 36(4): 10–15, 36–41.

Mitchell, Ronald B. 1996. "Compliance Theory: An Overview." In James Cameron, Jacob Werksman, and Peter Roderick, eds., *Improving Compliance with International Environmental Law.* London: Earthscan Publications, 3–28.

Mitchell, Ronald B. n.d. "Quantitative Analysis in International Environmental Politics: Toward a Theory of Relative Effectiveness." In Arild Underdal and Oran R. Young, eds., *Regime Consequences: Methodological Challenges and Research Strategies.* Unpublished ms.

Munton, Don 1998. "Dispelling the Myths of the Acid Rain Story." *Environment,* 40(6): 4–14.

National Marine Fisheries Service. 1997. "Bering Sea Ecosystem—A Call to Action." Draft white paper, September 21.

National Research Council. 1996. *The Bering Sea Ecosystem.* Washington, DC: National Academy Press.

National Research Council. 1999a. *Sharing the Fish: Toward a National Policy on Individual Fishing Quotas.* Washington, DC: National Academy Press.

National Research Council. 1999b. *The Community Development Quota Program in Alaska and Lessons for the Western Pacific.* Washington, DC: National Academy Press.

National Research Council. 1999c. *Human Dimensions of Global Environmental Change: Research Pathways for the Next Decade.* Washington, DC: National Academy Press.

Norse, Elliott A., ed. 1993. *Global Marine Biological Diversity: A Strategy for Building Conservation into Decision Making.* Washington, DC: Island Press.

North, Douglass C. 1990. *Institutions, Institutional Change and Economic Performance.* Cambridge: Cambridge University Press.

Oberthür, Sebastian. 1999. "Linkages between the Montreal and Kyoto Protocols." Paper prepared for a UNU conference on Synergies and Coordination between Multilateral Environmental Agreements.

Olson, Mancur, Jr. 1965. *The Logic of Collective Action.* Cambridge: Harvard University Press.

Onuf, Nicholas Greenwood. 1989. *World of Our Making: Rules and Rule in Social Theory and International Relations.* Columbia: University of South Carolina Press.

Opschoor, Johannes B. and R. Kerry Turner. 1994. *Economic Incentives and Environmental Policies: Principles and Practice*. Dordrecht: Kluwer Academic Press.

Osherenko, Gail. 1988. "Can Comanagement Save Arctic Wildlife?" *Environment*, 20(6): 6–13, 29–34.

Osherenko, Gail. 1995. "Property Rights and Transformation in Russia: Institutional Change in the Far North." *Europe-Asia Studies*, 47: 1077–1108.

Osherenko, Gail and Oran R. Young. 1989. *The Age of the Arctic: Hot Conflicts and Cold Realities*. Cambridge: Cambridge University Press.

Ostrom, Elinor. 1990. *Governing the Commons: The Evolution of Institutions for Collective Action*. Cambridge: Cambridge University Press.

Ostrom, Elinor. 1998. "A Behavioral Approach to Rational Choice Theory of Collective Action." *American Political Science Review*, 92: 1–22.

Ostrom, Elinor, Joanna Burger, Christopher B. Field, Richard B. Norgaard, and David Policansky. 1999. "Revisiting the Commons: Local Lessons, Global Challenges." *Science*, 284: 278–282.

Oye, Kenneth A., ed. 1986. *Cooperation under Anarchy*. Princeton, NJ: Princeton University Press.

Parson, Edward and Owen Greene. 1995. "The Complex Chemistry of the International Ozone Agreements." *Environment*, 37(3): 16–20, 35–43.

Peluso, Nancy Lee. 1992. *Rich Forests, Poor People: Resource Control and Resistance in Java*. Berkeley: University of California Press.

Pinkerton, Evelyn, ed. 1989. *Co-operative Management of Local Fisheries*. Vancouver: University of British Columbia Press.

Ponting, Clive. 1992. *A Green History of the World*. New York: St. Martin's Press.

Powell, Walter W. and Paul J. DiMaggio, eds. 1991. *The New Institutionalism in Organizational Analysis*. Chicago: University of Chicago Press.

Pressman, Jeffrey and Aaron Wildavsky. 1973. *Implementation*. Berkeley: University of California Press.

Pritchard, Lowell, John Colding, Fikret Berkes, Uno Svedin, and Carl Folke. 1998. "The Problem of Fit between Ecosystems and Institutions." IHDP working paper no. 2. Bonn: IHDP.

Putnam, Robert D. 1988. "Diplomacy and Domestic Politics: The Logic of Two-Level Games." *International Organization*, 42: 427–460.

Ragin, Charles. 2000. *Fuzzy-Set Social Science*. Chicago: University of Chicago Press.

Raustiala, Kal and David G. Victor. 1996. "Biodiversity since Rio: The Future of the Convention on Biological Diversity." *Environment*, 38(4): 16–20, 37–45.

Reinicke, Wolfgang H. 1998. *Global Public Policy: Governing without Government*. Washington, DC: Brookings Institution Press.

Risse, Thomas. 2000. " 'Let's Argue!': Communicative Action in World Politics." *International Organization,* 54: 1–39.

Rittberger, Volker, ed. 1993. *Regime Theory and International Relations.* Oxford: Clarendon Press.

Roan, Sharon. 1989. *Ozone Crisis: The Fifteen-Year Evolution of a Sudden Global Emergency.* New York: John Wiley & Sons.

Rose, Carol. 2002. "Common Property, Regulatory Property, and Environmental Protection: Comparing Common Pool Resources to Tradable Environmental Allowances." In Thomas Dietz, Nives Dolsak, Elinor Ostrom, and Paul C. Stern, eds., *The Drama of the Commons: Institutions for Managing the Commons.* Washington, DC: National Academy Press.

Ruggie, John Gerard. 1983. "International Regimes, Transactions, and Change: Embedded Liberalism in the Postwar Economic Order." In Stephen D. Krasner, ed., *International Regimes.* Ithaca: Cornell University Press, 195–231.

Rutherford, Malcolm. 1994. *Institutions in Economics: The Old and the New Institutionalism.* Cambridge: Cambridge University Press.

Sand, Peter H. 1999. "Carrots without Sticks? New Financial Mechanisms for Global Environmental Agreements." *Max Planck Yearbook of United Nations Law,* 3: 363–388.

Sand, Peter H. 2000. "The Precautionary Principle: A European Perspective." *Human and Ecological Risk Assessment,* 6: 445–458.

Sandler, Todd. 1997. *Global Challenges: An Approach to Environmental, Political and Economic Problems.* Cambridge: Cambridge University Press.

Schelling, Thomas C. 1960. *The Strategy of Conflict.* Cambridge: Harvard University Press.

Schelling, Thomas C. 1978. *Micromotives and Macrobehavior.* New York: W.W. Norton.

Scott, W. Richard. 1995. *Institutions and Organizations.* Thousand Oaks, CA: Sage Publications.

Scrivener, David. 1999. "Arctic Environmental Cooperation in Transition." *Polar Record,* 35: 51–58.

Sebenius, James K. 1983. "Negotiation Arithmetic: Adding and Subtracting Issues and Parties." *International Organization,* 37: 281–316.

Sebenius, James K. 1984. *Negotiating the Law of the Sea.* Cambridge: Harvard University Press.

Sherman, Kenneth. 1992. "Large Marine Ecosystems." In *Encyclopedia of Earth System Science.* Vol. 2. New York: Academic Press, 653–673.

Simon, Herbert A. 1957. *Models of Man: Social and Rational.* New York: John Wiley & Sons.

Singleton, Sara and Michael Taylor. 1992. "Common Property, Collective Action, and Community." *Journal of Theoretical Politics,* 4: 309–324.

Skjærseth, Jon Birger. 2000. *The Making and Implementation of North Sea Pollution Commitments: Institutions, Rationality and Norms.* Manchester: University of Manchester Press.

Small, George L. 1971. *The Blue Whale.* New York: Columbia University Press.

Snidal, Duncan. 1985. "The Limits of Hegemonic Stability Theory." *International Organization,* 39: 579–614.

Social Learning Group. 2001. *Learning to Manage Global Environmental Risks: A Comparative History of Social Responses to Climate Change, Ozone Depletion, and Acid Precipitation.* Cambridge: MIT Press.

Soroos, Marvin S. 1997. *The Endangered Atmosphere: Preserving a Global Commons.* Columbia: University of South Carolina Press.

Stegner, Wallace. 1954. *Beyond the Hundredth Meridian: John Wesley Powell and the Second Opening of the West.* Boston: Houghton-Mifflin.

Stein, Arthur A. 1982. "Coordination and Collaboration Regimes in an Anarchic World." *International Organization,* 36: 299–324.

Stigler, George. 1975. *The Citizen and the State.* Chicago: University of Chicago Press.

Stokke, Olav Schram, ed. 2001a. *Governing High Seas Fisheries: The Interplay of Global and Regional Regimes.* Oxford: Oxford University Press.

Stokke, Olav Schram. 2001b. "The Loophole of the Barents Sea Fisheries Regime." In Olav Schram Stokke, ed., *Governing High Sea Fisheries: The Interplay of Global and Regional Regimes.* London: Oxford University Press, 273–301.

Stokke, Olav Schram and Davor Vidas, eds. 1996. *Governing the Antarctic: The Effectiveness and Legitimacy of the Antarctic Treaty System.* Cambridge: Cambridge University Press.

Stokke, Olav Schram, Lee G. Anderson, and Natalia Mirovitskaya. 1999. "The Barents Sea Fisheries." In Oran R. Young, ed., *The Effectiveness of International Environmental Regimes: Causal Connections and Behavioral Pathways.* Cambridge: MIT Press, 91–154.

Strange, Susan. 1983. "*Cave! Hic dragones:* A Critique of Regime Analysis." In Stephen D. Krasner, ed., *International Regimes.* Ithaca: Cornell University Press, 337–354.

Susskind, Lawrence E. 1994. *Environmental Diplomacy: Negotiating More Effective Global Agreements.* Oxford: Oxford University Press.

Svensson, Tom G. 1997. *The Sami and Their Land.* Oslo: Novus forlag.

Tetlock, Philip E. and Aaron Belkin, eds. 1996. *Counterfactual Thought Experiments in World Politics: Logical, Methodological, and Psychological Perspectives.* Princeton, NJ: Princeton University Press.

Tietenberg, Thomas. 2002. "The Tradable Permits Approach to Protecting the Commons: What Have We Learned?" In Thomas Dietz, Nives Dolsak, Elinor Ostrom, and Paul C. Stern, eds., *The Drama of the Commons: Institutions for Managing the Commons.* Washington, DC: National Academy Press.

Tolba, Mostafa K. with Iwona Rummel-Bulska. 1998. *Global Environmental Diplomacy: Negotiating Environmental Agreements for the World, 1973–1992.* Cambridge: MIT Press.

Tullock, Gordon. 1989. *The Economics of Special Privilege and Rent Seeking.* Boston: Kluwer Academic Publishers.

Turner, B. L., W. C. Clark, R. C. Kates, J. Richards, J. Mathews, and W. Meyer, eds. 1990. *The Earth as Transformed by Human Action.* Cambridge: Cambridge University Press.

Underdal, Arild. 1980. *The Politics of International Fisheries Management: The Case of the Northeast Atlantic.* Oslo: Universitetsforlag.

Underdal, Arild, ed. 1998. *The Politics of International Environmental Management.* Dordrecht: Kluwer Academic Publishers.

Underdal, Arild. 1999. "Methodological Challenges in the Study of Regime Effectiveness." Paper prepared for the workshop on the Study of Regime Consequences, Oslo.

Underdal, Arild and Kenneth Hanf, eds. 2000. *International Environmental Agreements and Domestic Politics: The Case of Acid Rain.* Aldershot: Ashgate Publishing.

Underdal, Arild and Oran R. Young, eds. n.d. *Regime Consequences: Methodological Challenges and Research Strategies.* Unpublished ms.

Victor, David G. and Julian E. Salt. 1994. "From Rio to Berlin: Managing Climate Change." *Environment,* 36(10): 6–15, 25–32.

Victor, David G., Kal Raustiala, and Eugene B. Skolnikoff, eds. 1998. *The Implementation and Effectiveness of International Environmental Commitments.* Cambridge: MIT Press.

Vitousek, Peter, Harold Monney, Jane Lubchenko, and Jerry Melillo. 1997. "Human Domination of the Earth's Ecosystems." *Science,* 277: 494–499.

von Moltke, Konrad. 1997. "Institutional Interactions: The Structure of Regimes for Trade and the Environment." In Oran R. Young, ed., *Global Governance: Drawing Insights from the Environmental Experience.* Cambridge: MIT Press, 247–272.

Walton, Richard and Robert B. McKersie. 1965. *A Behavioral Theory of Labor Negotiations.* New York: McGraw-Hill.

Wapner, Paul. 1996. *Environmental Activism and World Civic Politics.* Albany: SUNY Press.

Wapner, Paul. 1997. "Governance in Global Civil Society." In Oran R. Young, ed., *Global Governance: Drawing Insights from the Environmental Experience.* Cambridge: MIT Press, 65–84.

Warner, William. 1983. *Distant Water: The Fate of the North Atlantic Fisherman.* Boston: Little, Brown.

Weiss, Edith Brown and Harold K. Jacobson, eds. 1998. *Engaging Countries: Strengthening Compliance with International Environmental Agreements.* Cambridge: MIT Press.

Wendt, Alexander. 1987. "The Agent-Structure Problem in International Relations Theory." *International Organization,* 41: 335–370.

Wendt, Alexander. 1992. "Anarchy Is What States Make of It: The Social Construction of Power Politics." *International Organization,* 46: 391–425.

Wendt, Alexander. 1999. *Social Theory of International Politics.* Cambridge: Cambridge University Press.

Wenzel, George. 1991. *Animal Rights, Human Rights: Ecology, Economy, and Ideology in the Canadian Arctic.* Toronto: University of Toronto Press.

White, Lynn, Jr. 1967. "The Historical Roots of Our Ecologic Crisis." *Science,* 155: 1203–1207.

Wilson, James A., James M. Acheson, Mark Metcalfe, and Peter Kleban. 1994. "Chaos, Complexity, and Community Management of Fisheries." *Marine Policy,* 18: 291–305.

World Wildlife Fund (WWF) and the Nature Conservancy of Alaska (TNC). 1999. *Ecoregion-Based Conservation in the Bering Sea.* Anchorage: WWF/TNC.

Worster, Donald. 1979. *Dust Bowl: The Southern Plains in the 1930s.* New York: Oxford University Press.

Young, Oran R. 1979. *Compliance and Public Authority: A Theory with International Applications.* Baltimore: Johns Hopkins University Press.

Young, Oran R. 1982. *Resource Regimes: Natural Resources and Social Institutions.* Berkeley: University of California Press.

Young, Oran R. 1983. "Fishing by Permit: Restricted Common Property in Practice." *Ocean Development and International Law,* 13: 121–170.

Young, Oran R. 1989. *International Cooperation: Building Regimes for Natural Resources and the Environment.* Ithaca: Cornell University Press.

Young, Oran R. 1994a. *International Governance: Protecting the Environment in a Stateless Society.* Ithaca: Cornell University Press.

Young, Oran R. 1994b. "The Problem of Scale in Human/Environment Relations." *Journal of Theoretical Politics,* 6: 429–447.

Young, Oran R. 1996. "Institutional Linkages in International Society." *Global Governance,* 2: 1–24.

Young, Oran R. 1998. *Creating Regimes: Arctic Accords and International Governance.* Ithaca: Cornell University Press.

Young, Oran R., ed. 1999a. *The Effectiveness of International Environmental Regimes: Causal Connections and Behavioral Mechanisms.* Cambridge: MIT Press.

Young, Oran R. 1999b. "Hitting the Mark: Why Are Some International Environmental Agreements More Successful than Others?" *Environment,* 41(8): 20–29.

Young, Oran R. 1999c. *Governance in World Affairs.* Ithaca: Cornell University Press.

Young, Oran R. 2001a. "The Behavioral Effects of Environmental Regimes: Collective-Action vs. Social-Practice Models." *International Environmental Agreements,* 1: 9–29

Young, Oran R. 2001b. "Inferences and Indices: Evaluating the Effectiveness of International Environmental Regimes." *Global Environmental Politics,* 1: 99–121.

Young, Oran R. and Gail Osherenko, eds. 1993. *Polar Politics: Creating International Environmental Regimes.* Ithaca: Cornell University Press.

Young, Oran R., Arun Agrawal, Leslie A. King, Peter H. Sand, Arild Underdal, and Merrilyn Wasson. 1999. "Institutional Dimensions of Global Environmental Change (IDGEC) Science Plan." IHDP report no. 9. Bonn: IHDP.

Zürn, Michael. 1999. "The State in the Post-National Constellation—Societal Denationalization and Multi-Level Governance." Unpublished essay. University of Bremen.

Index